VLSI Electronics
Microstructure Science

Volume 2

Contributors

Dennis R. Best

P. A. Blakey

W. F. Brinkman

R. W. Brodersen

J. R. East

D. K. Ferry

G. I. Haddad

K. Hess

Jacob Kline

Graydon B. Larrabee

Harold Levine

Robin Mallon

John Marley

Alan R. Reinberg

Alfred J. Stein

P. Vogl

VLSI Electronics
Microstructure Science

Volume 2

Edited by

Norman G. Einspruch

School of Engineering and Architecture
University of Miami
Coral Gables, Florida

1981

ACADEMIC PRESS

A Subsidiary of Harcourt Brace Jovanovich, Publishers

New York   London   Toronto   Sydney   San Francisco

6372 6051

CHEMISTRY

ACADEMIC PRESS, INC.
111 Fifth Avenue, New York, New York 10003

United Kingdom Edition published by
ACADEMIC PRESS, INC. (LONDON) LTD.
24/28 Oval Road, London NW1 7DX

Library of Congress Cataloging in Publication Data
Main entry under title:

VLSI electronics: Microstructure science.

   Includes bibliographical references and index.
   1. Integrated circuits--Very large scale
integration. I. Einspruch, Norman G.
TK7874.V56   621.381'73   81-2877
ISBN 0-12-234102-3 (v. 2)   AACR2

PRINTED IN THE UNITED STATES OF AMERICA

81 82 83 84   9 8 7 6 5 4 3 2 1

To my parents
Adolph and Mala

Contents

Chapter 8 The Impact of VLSI on Military Systems: A Supplier's Viewpoint

Dennis R. Best

Chapter 9 The Impact of VLSI on the Automobile of Tomorrow

Alfred J. Stein, John Marley, and Robin Mallon

Chapter 10 Very Large Scale Protection for VLSI?

Harold Levine

List of Contributors

Numbers in parentheses indicate the pages on which the authors' contributions begin.

Dennis R. Best (229), Microelectronic Circuit Development Laboratory, Equipment Group, Texas Instruments Incorporated, Dallas, Texas 75265

P. A. Blakey (105), Electron Physics Laboratory, Department of Electrical and Computer Engineering, University of Michigan, Ann Arbor, Michigan 48109

W. F. Brinkman (149), Bell Laboratories, Murray Hill, New Jersey 07974

R. W. Brodersen (189), Department of Electrical Engineering and Computer Sciences, University of California, Berkeley, California 94720

J. R. East (105), Electron Physics Laboratory, Department of Electrical and Computer Engineering, University of Michigan, Ann Arbor, Michigan 48109

D. K. Ferry (67), Department of Electrical Engineering, Colorado State University, Fort Collins, Colorado 80523

G. I. Haddad (105), Electron Physics Laboratory, Department of Electrical and Computer Engineering, University of Michigan, Ann Arbor, Michigan 48109

K. Hess (67), Department of Electrical Engineering and Coordinated Science Laboratory, University of Illinois, Urbana, Illinois 61801

Jacob Kline (171), Department of Biomedical Engineering, University of Miami, Coral Gables, Florida 33124

Graydon B. Larrabee (37), Texas Instruments Incorporated, Dallas, Texas 75265

Harold Levine (299), Sigalos and Levine, Dallas, Texas 75201

Robin Mallon (273), Motorola Semiconductors Japan Ltd., Tokyo 106, Japan

John Marley (273), Motorola Semiconductors Japan Ltd., Tokyo 106, Japan

Alan R. Reinberg (1), Optical Group Research, The Perkin–Elmer Corporation, Norwalk, Connecticut 06856

Alfred J. Stein (273), Semiconductor Group, Motorola Incorporated, Phoenix, Arizona 85062

P. Vogl (67), Institute for Theoretical Physics, University of Graz, Graz, Austria

Preface

Civilization has passed the threshold of the second industrial revolution. The first industrial revolution, which was based upon the steam engine, enabled man to multiply his physical capability to do work. The second industrial revolution, which is based upon semiconductor electronics, is enabling man to multiply his intellectual capabilities. VLSI (Very Large Scale Integration) electronics, the most advanced state of semiconductor electronics, represents a remarkable application of scientific knowledge to the requirements of technology. This treatise is published in recognition of the need for a comprehensive exposition that describes the state of this science and technology and that assesses trends for the future of VLSI electronics and the scientific base that supports its development.

These volumes are addressed to scientists and engineers who wish to become familiar with this rapidly developing field, basic researchers interested in the physics and chemistry of materials and processes, device designers concerned with the fundamental character of and limitations to device performance, systems architects who will be charged with tying VLSI circuits together, and engineers concerned with utilization of VLSI circuits in specific areas of application.

This treatise includes subjects that range from microscopic aspects of materials behavior and device performance—through the technologies that are incorporated in the fabrication of VLSI circuits—to the comprehension of VLSI in systems applications.

The volumes are organized as a coherent series of stand-alone chapters, each prepared by a recognized authority. The chapters are written so that specific topics of interest can be read and digested without regard to chapters that appear elsewhere in the sequence.

There is a general concern that the base of science that underlies integrated circuit technology has been depleted to a considerable extent and is in need of revitalization; this issue is addressed in the National Research Council (National Academy of Sciences/National Academy of Engineering) report entitled ''Microstructure Science, Engineering and Technology.'' It is hoped that this treatise will provide background and stimulus for further work on the physics and chemistry of structures that have dimensions that lie in the submicrometer domain and the use of these structures in serving the needs of humankind.

I wish to acknowledge the able assistance provided by my secretary, Mrs. Lola Goldberg, throughout this project, and the contribution of Academic Press in preparing the index.

Contents of Other Volumes

Chapter 1

Dry Processing for Fabrication of VLSI Devices

ALAN R. REINBERG

Optical Group Research
The Perkin–Elmer Corporation
Norwalk, Connecticut

I. INTRODUCTION

The process of changing a flat silicon wafer into a functioning electron device makes use of several different structure-forming techniques. The more commonly recognized are: chemical conversion of silicon into silicon oxide; film deposition by evaporation, sputtering, chemical vapor deposition (CVD), plasma-activated CVD, and mechanical processes such as spinning and spraying; diffusion; ion implantation; pulsed annealing; and selective removal of material by etching or lift-off. Except for mechanical deposition, which frequently involves the use of solvents, and selective removal processes, for which liquids are also required, most of the processes may be classified as "dry." The use of dry etch techniques has already become commonplace at some stages in the manufacture of state of the art integrated circuits and it is generally believed that the successful approach to the manufacture of micrometer and submicrometer devices will have to make extensive use of dry etching.

Although many forms of dry processing have been used for integrated circuit manufacture for some time, the term "dry processing" has come to mean, principally, the use of gas and vacuum processes for the selective removal of material by etching. The processes considered employ ordinarily inert gases that are activated by an electric discharge, forming a plasma consisting of electrons, atomic and molecular ions, photons, and molecular fragments or radicals. These new species react with materials to produce volatile products, which are then removed from the system. This subject constitutes the main body of this chapter. In it, we shall not attempt to give recipes and formulas for etching in a form that applies to any particular apparatus. Rather, an attempt is made to present direction and motivation for pursuing innovation in developing new processes and applying known ones to new situations involving microstructure formation.

During the past years there has been a proliferation of available machinery to accomplish this task in a production environment [1]. Many reviews of plasma and other dry etching techniques have begun to appear [2–5]. How-to-do-it application notes are available from several equipment manufacturers, who also sponsor periodic workshops. More significantly, there has been an increase in understanding of the complex physiochemical processes that produce the observed etching. However, in spite of the fact that all of the commonly encountered materials used in integrated circuit manufacture can be patterned using dry processing, the complexity of modern circuits and the current level of understanding of the processes will require considerable additional development to turn the art of dry etching into a science.

A. Need for Dry Processing

1. Geometry Control

The major motivation for increased usage of dry processing techniques for VLSI fabrication is the requirement for precise replication of lithography in the underlying layer. This is most readily accomplished with highly anisotropic etches. *Anisotropy* is defined as the vertical dimension change (etching rate) divided by the horizontal change [6]. Perfectly isotropic processes have an anisotropy of 1. Isotropic etching produces a feature profile that has a circular cross section. Small amounts of overetching, which are usually necessary to compensate for nonuniformities in the film properties and in the etching, produce large dimensional changes from the original resist pattern as shown in Fig. 1. For infinite anisotropy (zero horizontal dimension change), the pattern in the mask is reproduced precisely, so long as the mask is nonerodible [7]. Most common lithographic methods have some measurable slope to the photoresist profile and, as most organic masks evidence some erosion for common dry processing methods, the exact prediction of feature size is difficult. For a

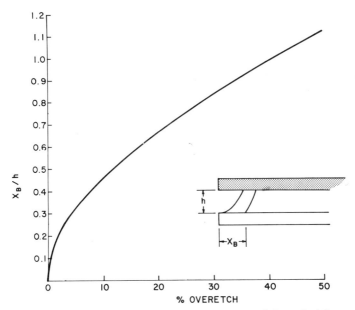

Fig. 1. Feature size variation as measured at the bottom of the etched feature for isotropic etching. Overetching is necessary to compensate for film and etch rate nonuniformities. Difficulties in controlling feature size can be compounded by increases in the lateral etch rate that occur after the film is etched through.

linear resist edge inclined an angle of θ from the horizontal, the change u in size from the initial position of the resist–surface contact is given as

$$u = 2h/\tan \theta, \tag{1}$$

where h is the amount of resist removed. A curved edge profile, which can result from plastic flow during baking, produces improved definition except for highly sloped resist. This comparison is shown in Fig. 2.

Resist geometry changes, like any other process-induced variations, are acceptable as long as they are uniform, controllable, and predictable across an entire wafer, permitting compensation in the mask. The fractional variation du/u in feature size of a resist pattern to resist removal only is obtained from Eq. (1) as

$$du/u = 2 \, dh/u \tan \theta. \tag{2}$$

A detailed discussion of geometry control including resist failure is given by King in Chapter 2 of Volume 1. He derives an expression for the variations due to the lithographic technique. The term dh is due to the *nonuniformity* in the resist removal caused by the etching process, assumed to be principally anisotropic. In most instances, the resist removal rate can be approximated as some fraction of the etch rate of the underlying film. Thus this term is directly related to the uniformity of the etching. It does not mean that the etched feature in the film will have the same variation

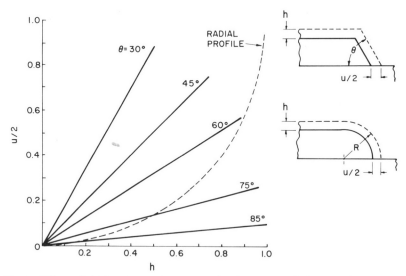

Fig. 2. Effect of initial resist profile on dimensional accuracy of a feature. Finite resist removal during etching of underlying film changes the position of the resist–film interface. Actual feature profile requires computer calculation.

but refers only to the final resist edge dimension. The actual shape and size of the etched feature can be obtained from a computer convolution of the film etch rate and anisotropy with the moving resist edge [8]. Viswanathan [9] has presented results for several special cases, a specimen of which is shown in Fig. 3. These examples assume anisotropic etching of the resist and the underlying film. Computer programs for generating profiles for arbitrary etching conditions are also available [10].

A value for edge acuity and feature slope can be obtained from an application of Eqs. (1) and (2) using estimates for the variations due to resist erosion and lithography. Assume a 1-μm-thick film, an initial resist of slope 70°, and a resist removal of 0.2 μm (out of an initial thickness of 1 μm) with a uniformity across the wafer of 15%, which is taken to represent the actual film etching uniformity. With these parameters the change u in geometry of the resist at the wafer surface is 0.15 μm and the variation is 0.02 μm. With erodible resists and realistic lithographic and etching methods, it appears that 1-μm lines can be readily patterned with a variation of less than 0.1 μm at the resist–film interface. Improvements in

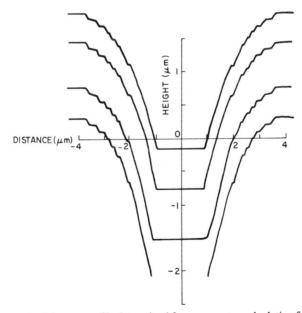

Fig. 3. Example of feature profile determined from computer calculation for an erodable resist mask. Careful control of resist and film etch rates can be useful for producing edges and contact windows with easy to cover contours while maintaining dimensional control. For directional etching using an erodable mask with $\theta = 60°$, the resist etch rate was 30 nm/min and the oxide etch rate was 40 nm/min. The etch profiles were computed times the percentage of 100 (normal) divided by 10, 50, 110, and 150. From Viswanathan [9].

process performance from this value are plausible but necessitate the use of nonerodible resist and highly anisotropic etches. A more complete discussion of this point is given in Chapter 2 by King in Volume 1. Of course, some amount of resist erosion is useful, even with anisotropic etches, if tapered walls and interconnects (vias) are desired to facilitate step coverage for the next layer of material. Additionally, there is evidence that vertical walls and right angles at the bottom of etched, single-crystal silicon can enhance nucleation of defects that degrade device performance.

It is possible to produce isolated features, specifically islands (mesas), by a number of special tricks. It is not, however, possible to produce vias or equal lines and spaces without specifying and controlling a smallest definable feature. Masks can be compensated to allow for isotropic etching effects but this is limited by the smallest definable feature size available from the lithography and by the thickness of etched film [6]. The compensated mask size necessary to produce a 1-μm feature in a 1-μm-thick film as a function of the etching anisotropy is shown in Fig. 4.

Some films used in device construction will scale in thickness with device geometry but others, such as interlevel insulation and some conductors, will not. This creates the need for highly anisotropic etches for the resulting high aspect ratios.

Not all dry processing methods are anisotropic. The conditions and systems in which other profiles have been observed are illustrated in Fig. 5. The mechanisms responsible for some of the observations are rea-

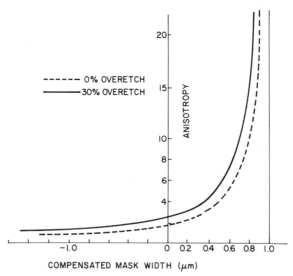

Fig. 4. Relation between anisotropy and compensated mask width required to etch a 1-μm-wide pattern in a 1-μm-thick film. The dimension refers to the top of the film where it interfaces with the resist mask. From Parry and Rodde [6].

ETCH PROFILE	STRUCTURE	SYSTEM OBSERVED	REACTOR TYPE	MECHANISM
ISOTROPIC	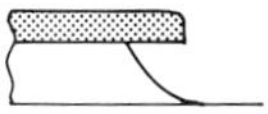	SILICON CF_4/O_2	ALL	CHEMICAL RESIDENCE TIME
VERTICAL ANISOTROPIC	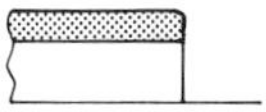	$SiO_2-C_X F_Y$ $Si-CCl_4$	PLANAR PLASMA, RIE	ION/ELECTRON ENHANCED SURF. CHEMISTRY
DIRECTIONAL ANISOTROPIC	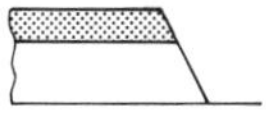	ALL	ION MILLING	PHYSICAL SPUTTERING
CRYSTALLINE ANISOTROPIC	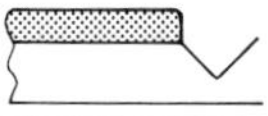	SILICON-CCl_4 KOH/H_2O	TUBULAR PLASMA, LIQUID PHASE	CHEMICAL
NEGATIVE UNDERCUT	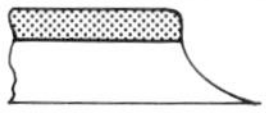	SILICON CF_4	TUBULAR PLASMA, PLANAR PLASMA	CHEMICAL ELECTRICAL ? QUENCHING ?
FINITE UNDERCUT	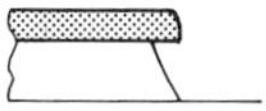	ALUMINUM MCl_X	PLANAR PLASMA	CHEMICAL CATALYTIC ?
ACCELERATED UNDERCUT	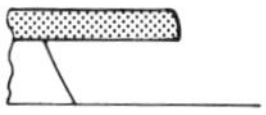	SILICON CF_4/O_2	ALL	CHEMICAL CONCENTRATION
REVERSE SLOPE	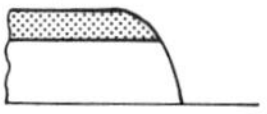	SILICON, OXIDE	RIE, PLANAR AFTERGLOW	RESIST FAILURE + VERTICAL ANISOTROPY ??

Fig. 5. Film profiles that have been observed for different etch processes. Many dry etching processes produce a combination of etching mechanisms that combine several of the simple profiles and produce complex edge shapes.

sonably well understood. Isotropic and accelerated undercuts are chemical in origin. The latter occur when a rapidly etching layer suddenly clears and exposes a slowly etching film. The amount of exposed film area of the fast layer is then suddenly reduced to a small fraction of its original value, causing a large increase in etch rate as a result of decreased loading [11,21]. Polymer resists can produce some feature size control that extends beyond simple masking. The negative undercut initially reported by Abe [12] has not been explained but is possibly also chemical in origin and may result from quenching of an active species by the masking resist layer. Depending on the etching method, the etch rate on a given wafer can be a function of the magnitude of the exposed area.

2. Yield Improvement

Because of the proprietary nature of any yield data, little information is available in the literature concerning the impact of dry processing. The usual means of quantifying yield improvement is the split lot, where one or more process variations are considered, "all other things being held equal." The problem lies, of course, in assuring that the latter condition is fulfilled or at least that significant differences are accounted for. In comparing dry and wet processes, for example, consideration should be given to the cleansing effect of fluids, which removes small particulate matter from the surface. Dust or dirt particles that would be removed by the flushing action of liquids may act as masks in the low-pressure gases used in plasma processes. Consequently, dry processes tend to require high-quality resist application and careful handling.

Dry processing may enhance yields because of improved geometry control due to anisotropic effects, reduction in resist adhesion problems or lifting, absence of the bubble formation common in liquid etching, and etchant piping and trapping. As previously indicated, however, dry processing can introduce its own failure mechanisms. Some problems are radiation damage, resist edge modification, and difficulties associated with anisotropic etching, which include stress-related problems associated with sharp corners, the inability of vertical etches to clear material out from reentrant regions, and difficulties in covering sharp steps.

In some cases the introduction of dry processes has involved a simple one for one replacement of a wet process without any additional changes. However, the previously used process was often optimized with respect to such factors as resist type, resist thickness, and bakes, and it was not a priori to be expected that the same optimum conditions would prevail for the dry process. Resist adhesion, for example, is usually not a problem in plasma etching, whereas resist removal rate (etching) may be important. In wet processing, exactly the opposite conditions usually prevail.

Dry processing is in principle compatible with a high degree of automation. This circumstance can be expected to lead to long-term yield improvements and will be discussed further. Additional gain can be expected in sequential processing, such as an etch followed by resist removal (ashing), which can be accomplished without additional handling. Finally, when failures or yield reduction are evidenced in a dry process step, it is often difficult to trace the source of the problem. Automation, if it is coupled with appropriate data logging, may be expected to aid in solution of process-induced yield loss.

3. Cost Impact

Early plasma etching apparatus, usually in the form of barrel reactors for ashing, were relatively simple in design, physically small for their wafer-handling capacity, and modest in cost. Modern etching machines frequently include complex mechanical, electrical, and vacuum systems incorporating significant automation and monitoring techniques; may cost up to $200,000; and occupy 20–30 ft^2 of floor space. It is likely that this trend toward sophistication will continue with significant initial equipment cost. The problem of high initial cost is compounded by the tendency in many production areas to dedicate a particular machine to a given process. A crude price comparison among the various dry etching processes suggests that, overall, plasma etching is the least expensive, with reactive ion etching next, and ion milling the most expensive. Typically, an automatic etcher can be expected to process between 60 and 100 wafers/hr depending on the nature of the film. On a three-shift basis, a machine could then perform about 5×10^5 wafer operations/yr. The economics of wafer fabrication will then dictate what are acceptable initial and operating costs for a particular apparatus.

It is generally accepted that material costs in dry processing offer substantial savings compared to wet etching. For many of the common etchants and other gases, this is true. However, there are exceptions for the more exotic materials. The most expensive etch gas currently used is most likely C_3F_8, which is employed in selective anisotropic oxide etching (to be discussed later). Because of short supply, this material currently sells at about \$100/lb. From simple stoichiometric considerations, one finds that 1 cm^3 of C_3F_8 should, in principle, be sufficient to process 2.0×10^5 cm^2 Å of SiO_2, or, in terms of 3-in. wafers, it can produce enough fluorine to etch one 3-in. wafer covered with 10,000 Å of oxide at approximately 50% surface coverage. The cost per wafer under these ideal conditions would be about 2¢. In practice, this usage factor may be exceeded by a factor of 5 or 10, giving a more realistic value of 10 to 20¢/3-in. wafer. This crude estimate should be considered an upper limit on gas costs in plasma etching and represents an extreme in raw material expense. The more common etch gases are about one-fifth the cost of C_3F_8.

4. Environmental Impact and Safety

Dry processing eliminates the need for storage and disposal of corrosive solutions but introduces other hazards. Under no circumstances does semiconductor device manufacture produce large-scale environmental hazards typical of other chemical-based industries. Circuit manufacturing can, however, produce dangerous conditions requiring careful considerations for personnel safety. While some reagents, particularly corrosives such as BCl_3, require special handling, the greatest dangers result from products formed in the plasma discharge itself. Chlorine-containing compounds can form phosgene ($COCl_2$), a highly toxic gas, when reacted with oxygen or upon exposing the products from a chlorine-containing discharge to atmospheric moisture. Various metal carbonyls, all of which are hazardous even in very small concentrations, can also be produced. Proper venting of pumps, purging of chambers, and appropriate fume hoods are essential. In addition, many of the species produced in the discharge react with pump oils, producing corrosive products that attack pumps. Contact by personnel with these products requires appropriate safety apparatus. Normal precautions for handling vacuum and high-voltage apparatus should also be employed.

Although several hundred thousand pounds per year of fluorocarbons are used in semiconductor dry processing, until now there have been no restrictions on the discharge of the products to the atmosphere. Should conditions require elimination of this effluent, small scrubbers could, in principle, be used if suitable solvent systems can be found.

5. Automation

Dry processing provides opportunity for automatic operation that is difficult to achieve by other means. Cassette-to-cassette operation or in-line operation and sequential process steps can impact yield by reducing the amount of handling. End point detection, automatically controlling etch time, can compensate for differences in film properties, as well as changes in etching conditions.

Automatic control of vacuum processes is well known in the semiconductor and thin-film industries. We may expect to see the same level of sophistication in dry etching apparatus. Operations that are amenable to automatic operation include: wafer loading and unloading; process sequencing; and vacuum cycling and real-time monitoring of process functions such as etch rates and film clearing (end point detection).

The "all dry factory" with wafers moving along tracks, under computer control, to different dry etching stations is a distinct possibility and may indeed be a reality in some advanced front-end fabrication facility. The major impact of this mode of operation is to be expected in high-volume production operations. At the low-to-medium throughput level, cassette-to-cassette etching machines will still be utilized but the control is more apt to rest with the process engineer than in some sophisticated computer program.

B. Alternatives to Dry Processing

1. Wet Processing

Wet chemical etching has received considerable attention as it has been the major form of material removal process in the fabrication of semiconductor devices. Its main failure is the lack of anisotropic etches, except in special cases involving single crystals. There is no known way in which amorphous materials can be removed anisotropically with current liquid etch technology.

2. Lift-Off

Ordinary subtractive pattern transfer uses the resist mask as a stencil, and material not protected by the mask is removed. Lift-off is an alternate technique in which a reversed image of the desired pattern is formed on the wafer with resist and the desired film deposited over it. The mask material may be an organic polymer, such as a photoresist, or some other easily dissolved layer. It is made thick enough so that the deposited film has very poor or incomplete step coverage. In this way a solvent or gas etchant can undercut those areas where the deposited film is not in intimate contact with the wafer surface. The undesired areas are thus "lifted off" the wafer. This technique can be used to produce extremely fine resolution. It is limited by the ability to pattern thick resist films with small geometries and by the restriction to deposited films that are compatible with the temperature limitations of the lifting layer. An automated lift-off process using plasma etching to remove the lifting layer would appear to have considerable possibility for patterning fine lines in materials such as aluminum and other metals.

3. Selective Nucleation

Additive processes differ from ordinary subtractive etching in that material is deposited in desired regions. They may be used to increase the thickness of a thinner, originally etched pattern or, alternatively, the entire pattern may be added. Electroplating of metals has been used to produce fine geometry masks for x-ray lithography. A totally additive process, without the use of liquids, has been demonstrated by the use of an electron beam to change the nature of a polymer film so that nucleation of the evaporated metal occurred only in specified regions. More commonly, photoactivated portions of a thin film of a tin-containing surface layer control the amount of palladium deposited on the surface by solution. The palladium coating in turn acts as a nucleation region for plating from an electroless copper plating solution. This process is used extensively in the preparation of circuit boards.

II. OUTLINE OF DRY ETCHING TECHNIQUES AND EQUIPMENT

There are several ways by which the various dry etching techniques can be classified. First, the applied parameters of the gas and discharge characteristics can be used. Important variables include pressure, gas composition, discharge geometry, frequency, system voltage and current, power and power density, and construction material. Alternatively, the physical characteristics of the active species may be described: What are they (ions, neutrals, fragments)? What are their origin, energy, directionality, lifetime, collision probability, and surface reactions? The latter set is of greater utility when attempting to explain some of the observations of etching characteristics. Unfortunately, in many important cases, the information is, as yet, unknown.

The forms of dry processing likely to impact device manufacture for the next decade or so include ion milling, plasma etching, and reactive ion etching (RIE). Ion milling is physical etching relying on momentum transfer to remove material in a manner similar to sputter etching. Plasma etching and reactive ion etching may also involve some momentum transfer, but chemical reactions producing volatile species that are pumped away are generally believed to be the dominant mechanism.

A. Process Effects

The operating characteristics of some of the dry processing methods are shown in Fig. 6. Other methods related to those illustrated include:

PROCESS		WET CHEMISTRY	PHYSICAL SPUTTERING & ION MILLING	PLASMA ETCHING			REACTIVE ION ETCHING
				BARREL	BARREL	DIODE	
WAFER LOCATION		IMMERSION	CATHODE	IMMERSION	TUNNEL	ANODE	CATHODE
SYSTEM SCHEMATIC							
ACTIVE SPECIES PRESENT	HIGH-ENERGY IONS		✓				✓
	LOW-ENERGY IONS			✓		✓	
	LONG-LIVED RADICALS			✓	✓	✓	✓
	SHORT-LIVED RADICALS			✓		✓	✓
	ATOMS & MOLECULES			✓	✓	✓	✓
CHEMICAL SELECTIVITY		EXCELLENT	POOR	GOOD	FAIR	GOOD	GOOD
ANISOTROPY OR LACK OF UNDERCUTTING (Determines Resolution)		POOR	EXCELLENT	VARIABLE	POOR	GOOD TO EXCELLENT	EXCELLENT TO GOOD
ETCH RATE		FAST	SLOW	FAST	FAST	FAST	SLOW
WAFER TEMP. CONTROL		EXCELLENT	GOOD	POOR	FAIR	GOOD	GOOD
MOS DIELECTRIC D'M'GE		NEGLIGIBLE	SIGNIFICANT(a)	SLIGHT (b)	NEGLIGIBLE	SLIGHT (b)	SIGNIFICANT(a)
REDEPOSITION		NEGLIGIBLE	SIGNIFICANT (c)	SLIGHT (d)	NEGLIGIBLE	SLIGHT (d)	SIGNIFICANT(c)
ETCH RATE UNIFORMITY		GOOD	GOOD	VARIABLE	GOOD	VARIABLE	GOOD

(a) Ion bombardment
(b) UV, electron bombardment, backscattered contaminants
(c) Backscattering of sputtered nonvolatiles
(d) Surface decomposition of radicals, backscattering of contaminants

Fig. 6. Qualitative comparison of various dry etch processes in their present state of advancement. Discussion of the features is given in the text.

downstream etchers similar to tunnel reactors, triode etchers that can span the gap between anode plasma etchers and reactive ion etching, and various alternate excitation modes, such as microwave [13], that may combine several effects. Depending on operating conditions, immersion and diode plasma etchers can also produce high-energy ions that bombard the wafer surface [14].

The key operational parameters of a dry etch process are selectivity, anisotropy, uniformity, and etch rate. Selectivity, which is the ability to etch a layer and stop on the underlying material, is a strong function of the chemical species produced in the discharge as well as the physical geometry of the apparatus.

There is general belief that anisotropy results from some form of directed energy normally incident on the surface [15]. If the incident particles are charged, they can be affected by variations in electric field direction near the surface of the workpiece [16]. Local field concentration can increase the ion or electron flux.

Etch rate is important, especially in large volume production. Rates are limited by the reactant supply [17,18], the available usable power for dissociation into active species, and the physical limitation due to resist or mask erosion. Also, for reasons not yet understood, some processes work better at slow rates while others do better when made faster. For example, it is often observed that oxide-to-silicon selectivity improves at low etch rates. The relative rates for the different processes are approximate. Actual values depend on many factors such as loading time, pump out time, and wafer load size, as well as actual film removal rate. Systems can be found for which these differences are less significant.

Dielectric damage, resulting in enhanced electron trapping in MOS devices, can be produced by penetrating and nonpenetrating radiation [19]. The only methods totally immune to these effects are the downstream etchers and, to a slightly lesser extent, the tunnel, or shielded, reactors.

Redeposition can occur from two different sources. First, material being etched can backscatter by gas collisions onto the surface. This can occur at pressures greater than about 0.1 Torr. Second, material from any other part of the chamber can be sputtered onto or react with the chemically active plasma and transported to the wafers. Some material is particularly susceptible to this type of transport. For example, gold can contaminate a chamber making it useless for etching-sensitive MOS devices [20] without drastic clean up procedures.

All the dry etch methods can be configured to produce uniform etching across a wafer for some processes. Nonuniformity can occur as batch-to-batch, wafer-to-wafer, or intrawafer effects. Batch-to-batch variations may be due to differences in material or changes that occur in the reactor.

Important process parameters include gas flow and concentration, pressure, discharge power, and temperature. In addition, some processes modify the surface of the reactor, coating them with a polymericlike material which can make them reactive surfaces that absorb atmospheric gases and interfere with the etch process. Precise control of these variations is difficult and requires sophisticated monitoring apparatus such as mass spectrometers.

Nonuniformity due to material variations between batches can be accommodated by end point detection. Wafer-to-wafer nonuniformity may occur in batch reactors because of variations in process environment such as positional variations in the etcher. In some processes, this can be minimized by careful adjustment of process parameters or selection of appropriate operating points. Generally, low pressures tend to reduce positional variations due to gas flow effects. For processes that are essentially chemical, concentration gradients due to loading effects [21] are probably the most significant cause of differences between wafers. They also account for some of the intrawafer variations seen in immersion barrel reactors. For ion-induced processes, variations in current density with position due to edge effects of electrodes and the proximity of alternate conduction paths can produce wafer-to-wafer as well as intrawafer variations. Some batch reactors attempt to minimize positional variations by moving wafers so that they sample different positions in the reactor. It is not entirely clear, however, whether in some processes the physical cause of nonuniformity stays in place with the reactor or moves with the wafer. An example of the latter is the case of radial etch gradients that occur across a wafer, independent of location. This often causes wafers to etch faster at the periphery and clear last at the center and is due to the difference in material between the wafer and support structure. For planar plasma etchers the radial nonuniformity may occasionally reverse its direction with the center of wafers etching faster than the edges. This phenomenon is interpreted as evidence that the supporting table is a sink for reactive components of the plasma.

These variations are significant for the manufacture of VLSI devices if the feature size and/or etching of underlying material is adversely affected. Figure 7 shows how uniformity is related to selectivity. In order to protect the underlying film, high selectivities are necessary if the system exhibits poor uniformity. At first it would seem that high selectivity and anisotropic etching are all that are required to give good patterning, independent of uniformity, and for some processes this is true. There are, however, a number of situations when high selectivities are not possible, for example, in etching an oxide film that terminates on another oxide layer of different composition.

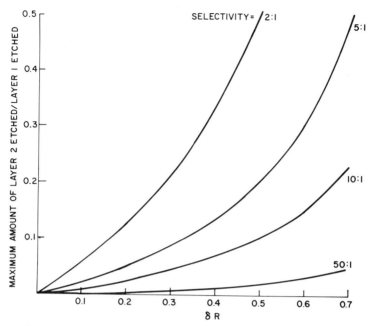

Fig. 7. Relation between selectivity and etch uniformity. Highly selective etches can relieve the uniformity requirements for an etching process.

An intrawafer nonuniformity effect that may be of significance as geometries shrink is a microscopic size effect. It is manifested as different-sized geometries etching at different rates. Smaller areas may etch faster than large ones or vice versa. Most likely it is due to a miniloading effect with the resist mask playing an active role in the etch process. If the mask is nonreactive to etching species, the concentration of etchant is greatest near the mask edge and near small geometries, which therefore etch faster. If the mask is a sink, the opposite situation prevails. If this problem is encountered, it may be necessary to ensure that critical lithographic steps do not contain grossly diverse feature sizes.

The problem of intrawafer nonuniformity will most likely increase with increased wafer size. It may make some of the newer, single-wafer processing systems more attractive. In this type of reactor, each wafer is processed individually, obviating effects associated with placement in a batch reactor. Additionally, if intrawafer nonuniformities exist, they can be adjusted for by adjusting gas flows, discharge plates, or other geometrical arrangements.

B. Plasma Etching

Generally speaking, plasma etching utilizes low-pressure glow discharges operating in the range from about 100 mTorr to several torr. Various frequencies, electrode configurations, and gas compositions are used to control etch rates, uniformity, selectivity, and profile of the etched film. The two main generic forms of plasma etchers are volume-loading and surface-loading machines or diode etchers.

1. Volume-Loading Etchers

Commonly referred to as barrel etchers, this form of etcher handles wafers stacked into a tube similar to a furnace tube but of varying dimensions. Excitation applied to the chamber creates a discharge that, depending on the chamber design, wafer loading, and excitation technique, produces reactive species. These species are usually concentrated close to the walls where the electrodes are placed and reach the wafer surface by diffusion. This, plus the strong dependence of etch rates of some materials on the amount of exposed area (loading effect), provides limited application of simple barrel or volume-loaded etchers for VLSI patterning. A major difficulty in performance is that combinations of etch mechanisms make contributions to the total etch rate, uniformity, profile, and selectivity. Some separation of process mechanisms can be accomplished in shielded reactors in which the discharge is confined between an annulus and the walls of the chamber. In these configurations, primary ionization is remote from the wafers while long-lived active components can reach them by free and forced diffusion.

In order to prevent the glow discharge plasma from penetrating into the region where the wafers are placed and negating the effectiveness of the shield, it is usually necessary to operate with high pressures or low power levels. By separating the discharge zone more completely from the wafer load, as in the downstream etcher described later, an additional degree of freedom can be obtained for optimizing the process.

The use of volume-loading etchers has almost been totally dedicated to fluorine-containing etchants for patterning silicon compounds. One exception has been the use of chlorinated species for chrome mask patterning. Generally, etch profiles tend to be isotropic, although some anisotropy is claimed for special etch gas systems [22].

If barrel reactors are used to etch a limited number of well-separated wafers, reasonably good results can be obtained by relying on end point detection to eliminate severe overetching and subsequent geometry loss. Process reproducibility is strongly dependent on appropriate temperature cycling of the chamber and several available units have special preheat cycles and temperature monitoring to assure that appropriate operating conditions have been reached. Poulsen and co-workers [23] have shown how careful temperature control can improve reproducibility of silicon oxide etching in shielded barrel reactors.

More recently, Doken and Miyata [24] have used a diffusion model to analyze etch rates of silicon in a tunnel reactor. They have shown that by adjusting the diffusion rates through appropriate baffle placement and careful wafer spacing uniformity can be improved.

Barrel reactors may be characterized as relatively long residence time reactors. Although values may differ among processes and specific equipment designs, gas residence times in the chambers are generally of the order of a few seconds. While diffusion coefficients for most of the species encountered are generally unknown, they can be expected, however, to be of the order of hundreds of square centimeters per second at 1 Torr. This supports the view that diffusion rather than forced convection (gas flow) is the dominant factor in determining etch uniformity in this system.

The long residence times for gases in barrel reactors result in high concentrations of long-lived radicals. Most of these contribute to chemical etching processes that are isotropic.

2. Diode Etchers

An alternate form of plasma reactor consists of a flat electrode on which the material being processed is placed. This geometry, sometimes referred to as a planar or diode etcher or surface-loading reactor, provides several advantages over the volume-loading barrel systems for micrometer and submicrometer patterning. Radial diffusion gradients on a given wafer are greatly reduced for most processes, although some variations may still exist because of the discontinuity in material at the periphery of the wafer. Positional variations in the reactor may be more bothersome and more dependent on gas flows than those in barrel reactors. Compared to the barrel reactors, diode etchers generally have higher gas flow rates and concomitantly shorter residence times of the reactant. Since wafers lie directly on a surface, temperature control is easier to accomplish than in barrel reactors where they are usually supported only at the periphery and float thermally. The most significant difference is that in diode etchers the wafer surfaces being etched are in direct contact with the plasma and, although nonuniformities due to positional variations in plasma intensity may occur, they are more readily accommodated and are considerably lower than those in volume-loading etchers.

Diode etchers are constructed in a wide range of sizes. Large-capacity machines holding many wafers are available in several configurations and now smaller systems that process one or a few wafers at a time have begun to appear. The trend to smaller wafer loads is based on the assumption that a process can be more readily optimized, selectivity, for example, over a small area.

In addition to the normal parameters (e.g., frequency, power, gas composition), planar reactors can be operated with wafers placed on either electrode, which can be at any potential available in the system. Commonly, material to be etched is placed on the electrode that is connected to ground (earth). When substrates are placed on the powered electrode and low pressures used (below 0.1 Torr), the system is generally referred to as reactive ion etching (RIE), although it is obvious that a full range of in-between conditions is possible.

C. Reactive Ion Etching

Reactive ion etching is a dry etch process in which discharge-produced ions directly or indirectly enhance the etch rates of selected materials. Compared to plasma etching, it generally requires lower pressures and higher ion energies. To some degree the name is unfortunate as it implies knowledge of the fact that ions react with the substrate material. Although in some instances this may be true, it is not at all clear that it is the only or even dominant mechanism. Other terms used to describe similar process technology are reactive sputter etching and ion-induced etching.

Irrespective of what it is called, the main features of the technique include pressures in the region of 10 to 100 mTorr, radiofrequency excitation (usually at 13.56 MHz although 27 MHz [25] has also been reported), and the use of diffusion or other high-vacuum pumps. (Plasma etching systems utilize mechanical pumps for the higher pressures at which they usually operate.) Wafers are placed on the cathode, which is the driven electrode.

As in rf sputtering, a positive dc bias [26] is created at the cathode because of the mobility difference between electrons and ions. This voltage, which is some fraction of the applied rf potential, accelerates ions to the workpiece. The removal of material may occur by several mechanisms, including sputtering, chemical reactions, and ion-induced chemistry.

Equipment construction and operation is very similar to that used in ordinary rf sputtering. The main difference being that the wafers are used as the target. In this respect it is closely related to rf sputter etching. Interest in RIE derives largely from its ability to produce very anisotropic etching of several materials. Use of very low pressures (compared to plasma etching) does not ensure that only vertical profiles are obtained. Chemical effects such as those active in plasma etching are also observed in RIE and produce undercutting of pattern stencils [27]. Reactive ion etching has been used to produce very impressive fine line geometries in many materials including silicon [25], silicon dioxide [28], and aluminum [29]. Sputter machinery specially designed for reactive ion etching in semiconductor manufacture is beginning to become available. With one exception (at this writing) they are manually loaded systems that open to atmosphere and, consequently, have limited throughput.

D. Downstream Etchers

Products from an electric discharge can be transported by a rapid flow system from the region where they are created to another region where wafers can be placed. Reactions of free radicals created in this manner have been studied for a number of years in what are often referred to as flowing afterglows, which implies the presence of reactive species giving rise to the luminescence and is common to many systems, particularly those containing atomic nitrogen.

Removal of the discharge zone from the etch region eliminates problems associated with resist degradation and device damage from high-energy particle bombardment. The downstream etchers are similar in this respect to tunnel barrel etchers. Wafers are, however, handled differently. The wafer loading region of a downstream etcher is usually planar as in the ordinary planar plasma etcher. Wafers sit on a heat sink surface, and the number of processed wafers depends on the size of the platten. By separating the discharge from the etching region, details of the process such as pressure and geometry can be optimized independently. The operating pressure of the discharge zone can be controlled independently of the pressure in the etching chamber. Generally, the former is high to improve power coupling and efficiency whereas the latter is low to improve uniformity and enhance transport of the reactive component from the discharge.

This approach has several advantages and disadvantages for VLSI processing. The two principal negative features are the limitation (as of this writing) to silicon and nonselective oxide and nitride etching and the totally isotropic character of the etch profiles. Compensating for these, however, are the excellent uniformity and control that can be achieved and the existence of a unique chemiluminescent end point indicator for (poly)silicon etching [30]. In addition, high rate etching can be achieved with essentially no resist degradation, making it attractive for use with thin, process-intolerant resists. The lack of anisotropy probably limits the method to 2-μm or larger geometries, except where exceptionally good uniformity and control can be achieved. In this case, 1-μm critical dimensions might be achieved in 0.5-μm-thick films primarily for polysilicon patterning. It is not clear, however, how this would be in any way preferable to anisotropic etching with the chlorine-containing etchants described in the next section.

E. Ion Milling and Sputter Etching

An energetic ion incident on a solid surface can transfer sufficient energy and momentum to an atom to eject it from the solid [31,32]. If the process is performed in what is traditionally a diode sputtering system, or a reactive ion etcher, it is generally referred to as *sputtering* [3]. On the other hand, *ion milling* implies a separation of the ion formation and acceleration system from the region in which the material being processed is placed. The latter system allows direct control over the angle of incidence with which the ions strike the surface. This additional degree of freedom is important in obtaining precise control of contours. The material removal rate and redeposition of sputtered material depend on the angle of incidence of the incoming ions. In the mill configuration this can be varied, while in the normal sputtering arrangement normal incidence is derived from the built-in electric field.

Nonreactive sputter etching, or ion milling, exhibits relatively poor selectivity between different materials and particularly against photoresist removal. This, plus the additional problem of radiation damage, has limited the use of these momentum transfer methods in silicon technology. Milling does have significant application in the manufacture of such devices as bubble memories, surface wave devices, integrated optics, and nonsilicon-active materials, where fine geometries are encountered as microwave FETs and the materials are not amenable to plasma etching. The advantages and disadvantages of ion etching are discussed in detail in the article by Melliar-Smith [3].

F. End Point Detection

It is desirable to terminate an etch process precisely at the time when the etching layer is penetrated. The actual utility of such end point determination for process control depends on how good an etch uniformity can be achieved. Dry processing provides several unique capabilities for monitoring etching. Mass [33], optical spectroscopy [34], and reflection spectroscopy have all been used successfully to determine etch end point. Electrical measurements [35] of the plasma have also been employed. Except for reflection, all of the techniques are integrating detectors in that in most implementations the observed signals are the result of average values of all materials exposed to the etch environment. Reflection spectroscopy, using a weak laser as a probe, determines the etching progress at a small point depending on the focus conditions. Of course, several beams can be used and appropriate averaging applied.

Emission spectroscopy of the plasma region of an etching apparatus provides a wealth of information concerning the components of the discharge. Table I lists some of the spectral lines that have been used to monitor changes in the composition of plasmas during etching of the common device materials. In some instances lines will increase as a particular layer etches through and no longer acts to suppress the concentration of the responsible species. Polysilicon and aluminum films are readily monitored by optical emission spectroscopy. On the other hand, silicon oxide, which often involves etching of a small total area on a wafer, is not a good candidate for averaging end point detection.

Mass spectroscopy is used to detect the etching product from various silicon compounds as well as to determine the overall condition of the vacuum systems in use. The major monitoring signal is that from SiF_3^+,

TABLE I

Wavelengths and Sources of Common Spectral Lines Used for
End Point Detection

Film	Etch gas	Line (μm)	Source
Silicon	CF_4	704	F* [b]
	$CF + O_2$		
	$CF_4 + O_2$	500–700	Chemiluminescence
	Downstream		(SiF_x)
Si_3N_4	$CF_4 + O_2$	674	N_2(?)
SiO_2	C_2F_6	483	CO
	C_3F_8		
Al	CCl_4	261[a]	AlCl
	BCl_3		
Resist	O_2	483	CO
		283	CO, OH
		298	
		309	

[a] Curtis and Brunner [35a].
[b] Excited fluorine atoms.

which represents the most detectable species from the ionization of silicon tetrafluoride (SiF_4), the main product from silicon reactions with fluorine. The most convenient way to sample these products is from the downstream effluent of the discharge. Care must be taken to minimize the response time of the system by the use of appropriately sized tubing. Sampling directly into the discharge by the use of small orifices is a useful experimental method but is difficult to accomplish in production machinery. Quadrupole mass spectrometers are commonly used for both experimental and production control. Usually the signals from SiF_3^+ (mass = 85) are sufficiently intense and remote from competing signals that simple detection schemes may be used. Faraday cup detection with external amplification can be used in place of the more expensive, easily poisoned, and fragile electron multipliers. Naturally, these systems are more limited in versatility.

III. SPECIFIC MATERIALS

A. Single-Crystal Silicon and Polysilicon

Dimensional control of polysilicon electrodes is critical in determining the performance of a wide class of integrated circuits. Silicon can be etched with either fluorine- or chlorine-containing gases. Some of the more common of these gases are listed in Table II. The etch mechanism can be varied from a totally chemical reaction in downstream etchers, which produce isotropic profiles, to vertical, anisotropic etching in planar reactors.

A unique chemiluminescence [30] observed in downstream etching may be used for end point detection if adequate uniformity is achieved. Because of reasons discussed previously, isotropic etching is limited to geometries of about 2 μm or larger. Anisotropic etching of silicon with fluorine etchants (CF_4) has been reported for planar, diode-configured etchers operating at low pressure [37]. These systems, however, have poor selectivity, making it difficult to stop on the thin oxide layers found in scaling to smaller dimensions. Chlorinated etchants will produce anisotropic profiles with good selectivity against oxide when used in planar reactors [25,38]. Both RIE and higher-pressure plasma conditions produce nearly vertical etching of silicon with CCL_4 and CCl_4 + Ar mixtures. A large variation in anisotropy is observed with different doping level of polycrystalline silicon, with heavily doped films exhibiting greatly enhanced isotropic etching [39,40]. Silicon-to-oxide etch ratios decrease in proportion to the number of fluorine atoms in the etchant. With doping

TABLE II

Silicon Etch Gases

Gas	System	Features
CF_4[a]	All	Isotropic but depends on condition
$CF_4 + O_2(1-10\%)$[b,c]	All	Enhanced silicon rate
$CF_4 + O_2(20-50\%)$[c]	Downstream	Isotropic, highly selective versus SiO_2
CCl_4, Cl_2[d]	Planar, RIE	Anisotropic
$SiF_4 + O_2$[e]	Barrel	Less isotropic than CF_4
$CF_3Br + C_2F_6$[f]	Planar	Anisotropic

[a] Flamm [42].
[b] Mogab et al. [36a].
[c] Horiike and Shibagaki [36b].
[d] Schwartz and Schaible [25].
[e] Boyd and Tang [22].
[f] Mogab [36c].

by ion implantation it is important that the implant be annealed prior to etching in chlorinated gases to avoid reentrant lateral etching of the high concentration and/or damaged layers. Alternatively, if possible, implant doping should be done after patterning. An additional advantage to the chlorine etching of silicon is the decreased loading effect.

Various faceting and trenching phenomena have been observed in chlorine etching of silicon similar to those seen in ion milling and sputtering [41]. The responsible mechanisms are not understood.

A critical comparison of polysilicon etch techniques has been published by Robb [42]. She concludes that even chemical etching in a downstream etcher may provide adequate control for polysilicon etching at geometries above 2.5 μm while some form of anisotropy is necessary for smaller critical dimensions.

A method for improving the anisotropy of silicon etching without the use of chlorinated etchants or very low pressures is by the use of intermediate sidewall oxidation. As described in the next section, silicon oxide etch rates increase upon exposure of the etching surface to ion (or electron) bombardment. In the limit of zero fluorine concentration, only those regions exposed to charged particle bombardment are etched. Thus if an isotropic etch is interrupted before complete removal of the polysilicon layer and exposed to an oxidizing atmosphere, a layer of silicon oxide forms on the etched surface. The undercut portion of the etched feature is shielded from the discharge and its oxide layer etches at a substantially reduced rate from the area exposed to the plasma.

B. Silicon Oxide

Oxide etching requirements for VLSI devices can be roughly divided into two categories: thick oxides (0.5–1.5 μm), often doped with phosphorous, and thin oxides (0.02–0.1 μm). Etching of small contact openings through thick oxides represents a severe test of dry etching technique. Silicon oxide can be etched with a wide variety of fluorine-containing gases in any of the reactor types described previously. Considerations for selecting a process include geometry control, selectivity, uniformity, radiation damage, and properties of the masking resist or stencil.

The factors outlined in Fig. 5 are a good guide to the etching of SiO_2 films. Tunnel etchers and, to a greater extent, downstream etchers produce isotropic, nonselective etching against silicon. They are, however, tolerant of organic resists and, hence, are useful for thick films. However, because of mask undercut, they cannot be relied on for high aspect ratio etching. Additionally, uniformity considerations may preclude their use in all but a few special situations. The main active species in these system is most likely fluorine atoms [36]. As in most chemical etch systems, rates increase with temperature, compounding the control problem. Of course, isotropic etching of oxides with fluorinated etchants can be used to provide taper to otherwise difficult to cover vertical steps and vias. However, to avoid excessive etching of underlying silicon layers, these process must be terminated prior to clearing the underlying film.

The most widely used approaches to dry oxide etching for VLSI applications are via parallel plate reactors. A nonplasma technique employing anhydrous HF has also been reported [43]. All of these techniques can produce the highly anisotropic etch profiles necessary for geometry control. The use of fluorinated etchants for this application is an area of current investigation, although the basic mechanism appears to be reasonably well understood [44]. A close relation between selective etching of silicon oxide and the production of fluorocarbon films has been observed [45]. Additionally, it is known that fluorocarbon films of varying structures can be made by plasma-enhanced deposition using several monomers [46]. While saturated fluorocarbons have the least tendency to form polymer films, unsaturated fluorocarbons such as C_2F_4 form polymers readily, as do those containing one or more hydrogen atoms such as CHF_3. Unsaturated fluorocarbons that have been used to produce selective etching of oxides are listed in Table III. The relation between polymer formation and selective oxide etching is illustrated in Fig. 8. The abscissa, labeled polymer formation, refers to any parameter that promotes the formation of polymers. Examples are increasing pressure, de-

TABLE III

SiO$_2$ Etch Gases

Gas	System	Features
CF$_4$[a]	Planar, RIE	Selectivity depends on conditions
CF$_4$ + O$_2$(1–10%)[b]	Planar, tube	Fast, but silicon much faster; variable anisotropy
CF$_4$ + O$_2$(20–50%)[c]	Downstream	Isotropic, temperature-dependent rate much slower than silicon
CF$_4$ + H$_2$[d,e]	Planar	Selective against silicon
C$_2$F$_6$	RIE	
C$_3$F$_8$		
CHF$_3$[d]		

[a] J. L. Mauer *et al.* [46a].
[b] Abe [46b].
[c] Flamm *et al.* [46c].
[d] Heinecke [46d].
[e] Coburn and Kay [45].

creasing temperature, and monomer concentration. Factors that decrease polymer formation are oxygen, probably free fluorine atoms, and ion bombardment of the surface.

Selectivity over silicon is primarily due to the oxygen in the SiO$_2$ providing an additional depolymerization mechanism. It is possible to find operating points using several of the etchants listed in Table III to obtain high selectivity etching of oxide versus silicon. Zero silicon etch rates (infinite selectivity) are observed for special cases, as are simultaneous oxide etching and polymer formation on silicon. Infinite selectivity conditions are hard to achieve with high uniformity but can have important

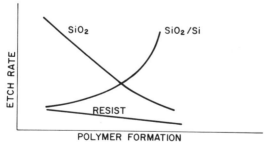

Fig. 8. Oxide and resist etch rates and oxide selectivity over silicon as a function of the tendency to form polymers. The abscissa represents any factor that increases the tendency to form polymers. Examples of these are: higher carbon to fluorine ratio in the etchant, higher partial pressure of carbon compounds, higher absolute pressure, and lower temperature.

applications when very shallow diffusions are present under the oxide being etched. Selectivities of 10:1 or greater have been reported for all the planar etching systems using many of the gases indicated in Table III. Although very high selectivity would be desirable, a factor of 10 to 15 appears to be readily achievable and sufficient for many applications.

Doping, primarily with phosphorous, can increase etch rates by as much as 50%, and it this has been proposed as the basis of a method to obtain tapered vias in otherwise anisotropic etching. It is not clear how such a process could be controlled sufficiently for VLSI geometries. Alternatively, controlled resist failure has also been suggested as a method for producing sloped via edges [47]. An alternate method is the use of isotropic etches to partially etch a film followed by anisotropic etching to limit the geometry at the bottom of the detail.

An interesting alternative to plasma or reactive ion etching of silicon oxides for VLSI application is the use of anhydrous HF without electrical activation [43]. In this process gaseous HF is absorbed into the photoresist and proceeds to etch the oxide that is covered by the resist. The resist moves through the oxide and etching stops when the oxide is penetrated down to bare silicon. This is a form of gel dissolution and provides extremely accurate reproduction of the resist image. The method has an advantage over anisotropic plasma etching in that the resist can act as self-aligned implantation mask.

C. Aluminum and Aluminum Alloys

Dry etching of aluminum metallizations may be expected to have a significant impact on the manufacture of VLSI devices. Aluminum widths of 1 μm have been demonstrated and production processes are becoming popular. The major tool for aluminum etching is the planar reactor, configured either as a plasma or reactive ion etcher [48,49]. The chemistry is based on interaction of chlorinated species to produce volatile aluminum chlorides. Under optimum conditions etch profiles are vertical, with little or no undercut, which is useful for geometry control but creates serious problems if the aluminum is applied over a previous steep geometry because of the difficulty in clearing out metal from the bottom of the step. In addition, the inability to etch under ledges causes immediate device failure.

Dry aluminum etching is a complex process involving several sequential steps [50]. First, the native oxide must be removed. It is generally believed that this requires some type of momentum transfer or sputtering process. Second, the aluminum itself must be etched without the production of the interfering polymer films that often occur in CCl_4 plasmas. Third, absorbed and adsorbed chlorine compounds must be removed prior to long-term exposure to humid air. If allowed to remain, they can lead to the formation of highly corrosive compounds that attack the metal and can completely destroy it. Finally, the resist must be removed. In addition, resist failure during etching can be troublesome. Considerable process latitude would result from the use of inorganic masking layers, but these are additional complications not readily introduced into the manufacturing process.

Pure aluminum is easier to etch than most alloys. However, copper and silicon doping are commonly used to reduce grain size and decrease current-induced metal transport. Since silicon can be etched with chlorinated plasmas, silicon-doped aluminum is more readily processed than copper. Often, however, the silicon and aluminum rates differ substantially, so that, after etching, some silicon residue remains on the wafer. This is readily removed in a short $CF_4 + O_2$ reaction of the type used to isotropically pattern silicon. Etching of copper-doped aluminum is more involved since copper does not readily form volatile chlorides, consequently copper residues that remain on the wafer must usually be removed by a wet chemical process.

Since chlorides also etch silicon and polysilicon, aluminum etching may lead to severe attack of an underlying silicon layer. It is observed, however, that the etch rate of aluminum is much less sensitive to power than that of silicon. Thus a programmed decrease in power may be used to discriminate against silicon after clearing the aluminum. Unfortunately, the anisotropy in aluminum etching may also be a function of power level so that simultaneous selectivity and anisotropy may be difficult to achieve.

An alternative to etching aluminum metallization patterns for fine geometries is the use of lift-off processing. This procedure requires the patterning of thick resists with special profiles, which is very difficult with conventional wet process. Dry processes may have some promise in this regard as will be discussed in Section III. G. Difficulties in automating the lift-off process limit its application to special cases.

D. Silicon Nitride

Used primarily as a diffusion mask, silicon nitride is readily etched in plasmas derived from CF_4, SiF_4, SF_6, and other fluorinated molecules. Both chemical- and ion-induced reactions are likely to be relevant. In downstream etching the nitride-to-oxide etch ratio is about $10:1$. The nitride-to-oxide ratios decrease as the system approaches a more ion-induced process. In most applications nitrides are deposited on oxide layers that must act as an etch stop. As devices are scaled and oxides become thinner, it is likely to become more difficult to do. Chlorine etch gases may play an increasingly important role, particularly as lateral dimension control becomes more important.

Since nitride films tend to be faily thin (600–1400 Å), undercutting is usually not severe. Very nonuniform etch processing as is often observed in barrel etchers may require significant overetching, resulting in unacceptable resolution over a large area. Generally the etching rate of silicon nitride is between that of silicon oxide and silicon and almost independent of the system employed. If nitrides etch faster than oxide, then silicon etches faster yet. Under conditions where nitrides etch slower than oxides, silicon etch rates are usually even slower.

E. Other Metals

Metals having volatile fluorides, such as molybdenum, tantalum, tungsten, and titanium, are readily etched in fluorine-containing etches derived from $CF_4 + O_2$. Since these materials have also been etched in barrel reactors [51], it is likely that some purely chemical etching occurs. When etched in planar reactors, most of these materials exhibit strong radial etch dependence indicative of a loading effect. The main use of these metals in present-day VLSI devices is as barriers. Consequently, they are thin, and geometry control is not critical. They can, however, if undercut sufficiently, develop poor profiles from a reliability point of view. Finally, since they form volatile fluorides, they do not etch selectively with respect to silicon, although high selectivity against oxide can be obtained.

Metal silicides such as $MoSi_2$ may begin to have a bigger role in VLSI devices and these too may be etched in $CF_4 + O_2$ discharges [52]. Unless new processes are devised that selectively remove them over silicon, it will be necessary to ensure that when used as conductors there is no necessity to stop on single-crystal silicon.

Metals that etch in chlorine-containing gases include chromium [53], gold [54], and platinum. Currently most dry etching of gold or platinum is by momentum transfer processes. Gold etching should not be attempted in any reactor used for general-purpose processing with fluoride etchants because residual gold products coat the interior of the chamber and contaminate subsequently processed material [20].

Chromium and vanadium are two examples of metals that have unusually volatile oxychlorides and, hence, are etched in mixtures of chlorine-containing gases and oxygen. Etch rates are a strong function of deposition conditions and techniques. Plasma etching of chromium photomasks is used extensively in industry to pattern high-resolution plates. Probably since the metal film on a photomask is thin (800 Å), little undercutting is observed in spite of large radial gradients in etch rate across a typical mask. Electric field gradients at the border between metals and insulators may be important. It is likely that plasma etching of photomasks will be necessary for optical lithography at VLSI dimensions in order to maintain geometry.

Several metals are particularly attractive for use as masks in ion milling of thicker films. Examples are aluminum and vanadium. Large differential milling rates can be achieved by adding small concentrations of oxygen to the milling gas.

F. III–V Compounds

Gallium arsenide is a potential candidate for very high speed devices which will require dry processing for patterning the necessary small geometries. It is an ideal substrate for supporting films that can be patterned with fluorine etchants since it is inert to these species. It can, however, be etched with chlorine-containing etchants, making the patterning of aluminum metallization difficult. There are no published reports of selective etches for any form of native oxide on GaAs that do not attack the underlying film. Insofar as these oxides serve only as gate dielectrics, it may be possible to design structures that do not require their selective etching. Finally, we note that dry processing in the form of plasma anodization may have an application for forming oxides on GaAs [55, 56].

G. Resists and Resist Systems

Any resist system must satisfy the following requirements: It must be able to produce patterns at the desired resolution. It must provide an edge profile consistent with the processing requirements and the desired geometry control. It must survive the etching process sufficiently well that it protects the underlying film at its weakest point. It must not introduce unwanted impurities into the etching system. Finally, it must be readily removable after processing.

Resist removal during dry processing is caused by physical bombardment, photochemical effects, chemical reactions with ions or radicals, and thermal degradation. The complexity of organic polymers precludes any but a few simple generalizations with respect to their usefulness for dry etching. Harada [57] has found that the plasma etch resistance of positive resists increases by the addition of radical scavengers and plastics' antioxidants. Resist survival can be increased by operating at lower temperatures and by subjecting to a high-temperature bake or other cross-linking environment prior to etching. High-temperature baking can, however, make the resist exceedingly difficult to remove after etching, particularly with chlorinated etchants. The survival of popular resists such as AZ 1350 depends dramatically on the particular process conditions. Selective etching of oxides, for examples, is generally also selective against resist and probably involves a similar polymer-forming tendency to that described in Section III.B.

VLSI requirements limit the choice of resists to those capable of fine-line resolution. For optical lithography, positive resist systems generally have higher resolving power than negative resists while at the same time providing thicker layers for improved protection.

Properties of several of the commoner e beam resists under $CF_4 + O_2$ etching conditions have been published by Jinno [58]. PBS [poly(butene sulfone)] is clearly the least resistant with several others, including PMMA [poly(methyl methacrylate)] and COP [copolymer of glycidyl methacrylate and ethyl acrylate], about the same. Since other conditions such as resist thickness and temperature resistance may dictate which material performs best, these criteria must be taken only as a crude guide to selection. Relative removal rates of a large number of polymeric materials by oxygen (and CF_4) plasmas have been presented in conference proceedings [59] and several factors that enhance or retard the removal rate have been identified. Those processes in which resist survival is likely to be a problem involve the etching of thick films such as interlevel insulation, thick field oxide, and metallization.

The probability that a polymer recording medium (resist) will satisfy all of the requirements for lithography and pattern transfer in one formulation is not very good. Highly sensitive resists may not have good survival in plasma conditions. Other requirements, such as the need for flat surfaces to provide high resolution, may dictate the use of multilevel composite resists, for example those described by Moran and Maydan [60]. A key feature in preparing this system is the use of reactive sputter etching of a thick organic film by oxygen to produce good linewidth control in very thick films. The major disadvantage of multilevel resists is that they require additional process steps.

Dry processing has recently been described for resist development to eliminate the use of organic solvents [61]. It has not, however, been shown to have any particular advantage for fine-line lithography over wet development when applied to organic resists. A dry-developed, inorganic,

Se–Ge photoresist has been shown to be capable of better than 1-μm resolution. Etching is performed with CF_4 in a commercial barrel plasma etcher [62].

IV. SUMMARY

A. Overall Status of Dry Processing

It is probably true that all of the processes necessary to make silicon integrated circuits with VLSI dimensions have been discovered. From the point of view that materials can be etched with appropriate selectivity, this statement is essentially correct. There are, however, many other factors that must be satisfied simultaneously with the ability to etch in order to have a viable production process. Figure 9 is a compilation of important factors and how the principal materials used in silicon VLSI manufacture relate to them. Those items that are in the author's opinion

	SILICON NITRIDE	SILICON OXIDE (thin)	POLY-SILICON	SILICON OXIDE (thick)	ALUMINUM
SELECTIVITY	X√	√	X	√	√
ANISOTROPY		X	√		√
RESIST SURVIVAL	X	X	X	√	√
UNIFORMITY	√	X	√		
ETCH RATE	X	X	X	√	X√
RADIATION EFFECTS		√	√	√	
PROCESS RELIABILITY	X		X	√	√

√ –IMPORTANT CONSIDERATION
X –READILY ACCOMPLISHED

Fig. 9. The present status of dry etch technology for materials used in the manufacture of integrated circuits. Items for important consideration (check marked) are those that arise in manufacture and that may impact device performance, yield, and reliability. Readily accomplished items (×) are those that can be realized in equipment that is available with well-known processes. Critical areas are those with check marks without crosses. Improvements in these features are important for viable large-scale manufacture of VLSI devices.

important considerations but not as yet readily accomplished may in fact be known by some readers to be practicable. They do not, however, represent well-known processes that are widespread manufacturing techniques. This lack of universality is due to the well-known secrecy of the semiconductor industry with respect to proprietary manufacturing techniques. The unavailability of appropriate process machinery is also a contributing factor.

It is likely that many of the known etch methods combined with some of the available machinery are adequate for 2-μm geometries. Some may even work well at the 1-μm level and the submicrometer levels but it is not clear that they are viable manufacturing techniques. This judgment will have to await the wider availability of lithographic tools at the micrometer and submicrometer feature sizes.

B. Present Trends

Considerable interest is being shown in anisotropic etching, careful control, and more monitoring of the entire etch process. Ion-dependent processes such as reactive ion etching are receiving much more attention and the first generation of commercial equipment to perform RIE is becoming available. The main goal is geometry control, but if this can be achieved by other techniques, they will be used if the overall cost is lower with substantially the same yield. Larger wafers are making batch reactors that etch several wafers at a time less attractive in comparison with one-at-a-time wafer etchers, several of which are being manufactured and sold. If process speed can be increased to make their throughput (per unit equipment cost) equivalent to batch reactors, they will see extensive application in manufacture for several processes.

Designers of new resist systems and lithographic tools are now asking if their processes are compatible with dry etching. Dry etching is changing from a replacement for wet etching to a dominant, controlling factor in process flow. It is likely that the dry etching methods and the results they produce will reach one step further back in device design and begin to impact the basic device architecture. Those organizations that are able to integrate this information into their device design structure will have a distinct competitive advantage.

C. Recommendations

In view of their importance for aluminum and anisotropic silicon etching, the chlorine systems are in need of detailed and expanded investigations. Process requirements in terms of anisotropy, selectivity, and uniformity have to be carefully evaluated and realistic expectations established that are consistent with anticipated dry processing techniques. By considering the limitations of the various dry process methods at a sufficiently early stage of device and process design, the total burden for obtaining a particular structure can be appropriately shared among several process steps instead of the often unrealistic attempt to make the etching step carry it all.

REFERENCES

1. P. S. Burggraaf, *Semicond. Int.* **2**, 49 (1979).
2. R. C. Booth and C. J. Heslop, *Thin Solid Films* **65**, 111 (1980).
3. C. W. Melliar-Smith, *J. Vac. Sci. Technol.* **13**, 1008 (1976).
4. A. R. Reinberg, "Etching for Pattern Definition" (M. J. Rand and H. G. Hughes, eds.), Softbound Symposium Series. Electrochemical Society, Princeton, New Jersey, 1976.
5. C. J. Mogab and C. W. Melliar-Smith, *in* "Thin Film Processes" (J. L. Vossen and W. Kern, eds.). Academic Press, New York, 1978.
6. P. D. Parry and A. F. Rodde, *Solid State Technol.* **22** (4), 125, (1979).
7. J. A. Bondur and R. G. Frieser, *in* Extended abstracts of the electrochemical Society Spring Meeting, St. Louis, Missouri, Abstr. #109. Electrochemical Society, Princeton, New Jersey, 1980.
8. J. L. Reynolds, A. R. Neurfuther, and W. G. Oldham, *J. Vac. Sci. Technol.* **16**, 1772 (1979).
9. N. S. Viswanathan, *J. Vac. Sci. Technol.* **16**, 338 1979.
10. University of California, Report on the design of a Simulator Program (Sample) for I. C. Fabrication, Memo #UCB/ERL M79/16. Univ. of California, Berkeley, California.
11. T. Enomoto, M. Denda, A. Yasuoka, and H. Nakata, *Jpn. J. Appl. Phys.* **18**, 155 (1979).
12. H. Abe, *Jpn. J. Appl. Phys.* **14**, 1825 (1975).
13. K. Suzuki, S. Okudaira, N. Sakudo, and I. Kanomata, *Jpn. J. Appl. Phys.* **16**, 1979 (1977).
14. J. L. Vossen, *J. Electrochem. Soc.* **126**, 319, (1979).
15. J. W. Coburn and H. F. Winters, *J. Vac. Sci. Technol.* **16**, 391 (1979).
16. K. Ukai and K. Hanazawa, *J. Vac. Sci. Technol.* **15**, 338 (1978).
17. B. N. Chapman and V. J. Minkiewicz, *J. Vac. Sci. Technol.* **15**, 329 (1978).
18. J. L. Mauer and J. S. Logan, *J. Vac. Sci. Technol.* **16**, 404, (1979).
19. D. J. DiMaria, L. M. Ephrath, and D. R. Young, *J. Appl. Phys.* **50**, 4015 (1979).
20. S. P. Murarka and C. J. Mojab, *J. Electron. Mater.* **8**, 763 (1979).
21. C. J. Mogab, *J. Electrochem. Soc.* **124**, 1262 (1977).
22. H. Boyd and M. S. Tang, *Solid State Technol.* **22**, (4) 133 (1979).
23. M. Brochu and R. G. Poulsen, *in* Extended abstracts of the Electrochemical Society Spring Meeting, Washington, D.C., Abstr. #143. Electrochemical Society, Princeton, New Jersey, 1976.
24. M. Doken and I. Miyata, *J. Electrochem. Soc.* **126**, 2235 (1979).
25. G. C. Schwartz and P. M. Schaible, *J. Vac. Sci. Technol.* **16**, 410 (1979).

26. H. S. Butler and G. S. Kino, *Phys. Fluids* **6**, 1346 (1963).
27. G. C. Schwartz, L. B. Rothman, and T. J. Schopen, *J. Electrochem. Soc.* **126**, 464 (1979).
28. H. W. Lehmann and R. Widmer, *Appl. Phys. Lett.* **32**, 163 (1978).
29. P. M. Schaible, W. C. Metzger, and J. P. Anderson, *J. Vac. Sci. Technol.* **15**, 334 (1978).
30. C. I. M. Benakker and R. P. J. Van de Poll, *Proc. Int. Symp. Plasma Chem., 4th, Zurich* (S. Veprek and J. Hertz, eds.), Vol. 1, pp. 125–1979. Univ. of Zurich, Zurich.
31. P. G. Gloersen and R. E. Lee, *in* Topics in Applied Physics." Springer-Verlag, Berlin and New York (in press).
32. G. K. Wehner and G. S. Anderson, *in* "Handbook of Thin Film Technology" (L. Maissel and R. Glang, eds.), McGraw-Hill, New York, 1970.
33. B. A. Raby, *J. Vac. Sci. Technol.* **15**, 205 (1978).
34. R. G. Frieser and J. Nogay, *Appl. Spectrosc.* **34**, 31 (1980).
35. K. Ukai and K. Hanazwa, *J. Vac. Sci. Technol.* **16**, 385 (1979).
35a. B. J. Curtis and H. J. Brunner, *J. Electrochem. Soc.* **125**, 829 (1978).
36. D. Flamm, *Solid State Technol.* **22** (4), 109 (1979).
36a. C. J. Mogab, A. C. Adams, and D. L. Flamm, *J. Appl. Phys.* **49**, 3796 (1978).
36b. Y. Horiike and M. Shibagaki, *Jpn. J. Appl. Phys. Suppl.* **15**, 13 (1976).
36c. C. J. Mogab, Paper presented at *Nat. Symp. Am. Voc. Soc., 26th, New York* (October 1979).
37. J. A. Bondur, *J. Vac. Sci. Technol.* **13**, 1023 (1976).
38. H. Kinoshita and K. Jinno, *Jpn. J. Appl. Phys.* **16**, 381 (1977).
39. H. B. Pogge, J. A. Bondur, and P. J, Burkhardt, *J. Electrochem. Soc.* **125**, 470C (1978).
40. G. C. Schwartz and L. B. Zielinski, Extended abstracts of the Electrochemical Society, Fall Meeting, Atlanta, Abstract #140. Electrochem Society, Princeton, New Jersey, 1977.
41. P. G. Gloersen, *J. Vac. Sci. Technol.* **12**, 28 (1975).
42. F. Robb, *Semicond. Int.* **2**, 60 (1979).
43. R. L. Bersin and R. F. Reichelderfer, *Solid State Technol.* **20** (4), 78 (1977).
44. J. W. Çoburn and H. Winters, *J. Vac. Sci. Technol.* **16**, 391 (1979).
45. J. W. Coburn and E. Kay, *IBM J. Res. Dev.* **23**, 33 (1979).
46. E. Kay, J. W. Coburn, and G. Kruppa, *Le Vide* No. 183, 89 (1976).
46a. J. L. Mauer, J. S. Logan, L. Zielenski, and G. Schwartz, *J. Vac. Sci. Technol.* **15**, 1734 (1978).
46b. H. Abe, *Jpn. J. Appl. Phys. Suppl.* **44**, 287 (1975).
46c. D. L. Flamm, C. J. Mogab, and F. Sklauer, *J. Appl. Phys.* **50**, 6211 (1979).
46d. R. Heinecke, *Solid State Electron.* **28**, 1145 (1975).
47. J. A. Bondur and H. A. Clark, *Solid State Technol.* **23** (4), 122 (1980).
48. K. Tokunaga and D. W. Hess, *J. Electrochem Soc.* **127**, 928 (1980).
49. P. M. Schaible, W. C. Metzger, and J. P. Anderson, *J. Vac. Sci. Technol.* **16**, 377 (1979).
50. D. K. Ranadive and D. L. Losee, *in* Extended abstracts of the Electrochemical Society Spring Meeting, St. Louis, Missouri, Abstr. #116. Electrochemical Society, Princeton, New Jersey, 1980.
51. K. Maeda and K. Fujino, *Denki Kagakuoyobi Kogyo Butsuri Kagako* **43**, 22, (1975).
52. T. P. Chow and A. J. Steckl, *in* Extended abstracts of the Electrochemical Society Spring Meeting, St. Louis, Missouri, Abstract #119. Electrochemical Society, Princeton, New Jersey, 1980.
53. H. Abe, *J. Jpn. Soc. Appl. Phys. Suppl.* **44**, 287 (1975).

54. C. B. Zarowin, *in* Extended abstracts of the Electrochemical Society Fall Meeting, October, Pittsburgh, Pa. Abstract #195. Electrochemical Society, Princeton, New Jersey, 1978.
55. R. P. H. Chang and J. J. Coleman, *Appl. Phys. Lett.* **32,** 332 (1978).
56. S. Gourrier and A. Mircea, *Thin Solid Films* **65,** 315 (1980).
57. K. Harada, *J. Electrochem. Soc.* **127, 491** (1980).
58. K. Jinno, *Jpn. J. Appl. Phys.* **17,** 1283 (1978).
59. G. N. Taylor and T. M. Wolf, *in* Photopolymers: Principles processes, and materials, *Proc. Conf. Soc. Plast. Eng., Inc., New York* (October 1979).
60. J. M. Moran and D. Maydan, *Bell Syst. Tech. J.* **58,** 1027 (1979).
61. J. N. Smith, H. G. Hughes, J. V. Keller, W. R. Goodner, and T. E. Wood, *Semicond. Int.* **2,** 41 (1979).
62. A. Yoshikawa, O. Ochi, and Y. Mizushima, *Appl. Phys. Lett.* **36** 107 (1980).

Chapter **2**

Materials Characterization for VLSI

GRAYDON B. LARRABEE

Texas Instruments Incorporated
Dallas, Texas

I. INTRODUCTION

The characterization of materials at small dimensions requires the concerted measurement of chemical, structural, and functional (e.g., electrical, magnetic) properties concomitant with sufficient understanding of their interaction to enable reproduction of the material. The measurement of these properties becomes increasingly difficult as (i) the form of the

37

material changes from bulk to film to surface, (ii) the physical size decreases from crystal to slice to chip to individual device element, and (iii) the elemental concentration decreases from major component ($>0.1\%$) to dopant ($1-1000$ ppma) to trace (<1 ppma). With the emergence of submicron technologies the three trends just described are happening simultaneously. Figure 1 illustrates the challenge faced both in device fabrication and, subsequently, in materials characterization. VLSI devices will be patterned at micron and submicron dimensions using ~ 1000-Å electron-beam lithography. In order to adequately image and analyze structures of this size, it will be necessary to have characterization $x-y$ spatial resolutions in the $10-100$-Å range. Further, it will be necessary to perform in-depth materials characterization with resolutions in the $20-30$-Å range. It is apparent that three-dimensional analysis in the x, y, and z planes will be an integral part of VLSI research and development in materials, processes, and devices.

This decrease in device feature size will also cause physical defects, e.g., dislocations, stacking faults, and precipitates of small dimensions to become more important in degrading yields, device performance, and device reliability. As will be shown later, Section IV.B.2, the decoration of a physical defect with clusters of impurity atoms as small as $50-100$ Å in diameter can result in electrical activity that degrades device performance and yield. This effect was observed in devices built with 5- to 7-μm design rules where CCD charge storage is ~ 100K electrons and DRAM cell storage of ~ 1M-5M electrons. Obviously submicron devices will move

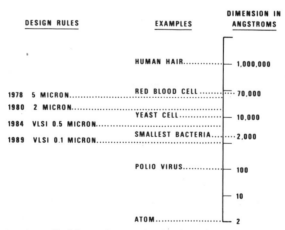

Fig. 1. Comparison of minimum feature size (design rules) on current and projected devices with structures produced in nature.

and store much smaller numbers of electrons and will be particularly vulnerable to this type of problem.

The role of submicron materials characterization then becomes clearer. It is essential that the interactive role of chemical impurities, physical defects, and their electrical activity or function be understood. In order to establish this understanding it is necessary to utilize a variety of complementary materials characterization techniques. The specific role of each technique, its strengths and weaknesses, must be clearly understood when dealing with submicron geometries associated with VLSI. Larrabee [1], Kane and Larrabee [2], and Evans and Blattner [3] have written comprehensive reviews of the capabilities of materials characterization techniques for surfaces and thin films. This chapter will examine characterization techniques for chemical analysis, surface and chemical imaging, physical defects, and functional properties with particular reference to VLSI materials, processes, and devices.

II. CHEMICAL ANALYSIS

The characterization of a material or a device for chemical impurities in the realm of VLSI technology requires more than a measurement of the average bulk concentration. It is important to know where the chemical entity is located $(x-y)$ and how it is distributed with depth into the matrix (z). The most frequently used probes to obtain this information are beams of electrons, ions, and x rays. Figure 2 shows the $x-y$ resolution capabilities of these three methods of excitation. Only electron beams are capable of being focused to less than 1 μm in diameter. Electron-beam excitation with beam diameters as small as ~ 2 Å is attainable with today's advanced characterization instrumentation.

There is an interesting dichotomy in the characterization of materials at small dimensions. It involves a fundamental concept in microvolume analysis, which will be interwined throughout this chapter, that high-sensitivity analysis and microvolume analysis are mutually exclusive. As will be seen later there are a number of reasons for this, but one major reason is shown in Fig. 3. As can be seen, when the excitation-beam diameter is decreased, there is a corresponding decrease in electron- or ion-beam current. This results in less excitation of the sample and thereby decreased sensitivity and poorer detection limits. The highest $x-y$ spatial resolution is obtained at the lowest beam currents where elemental sensitivity is the poorest.

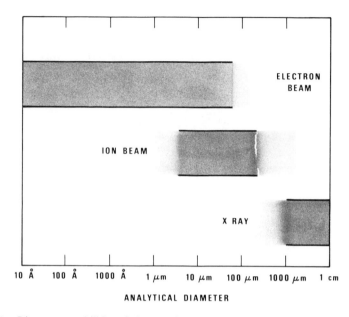

Fig. 2. Diameter capabilities of electron, ion, and x-ray characterization techniques.

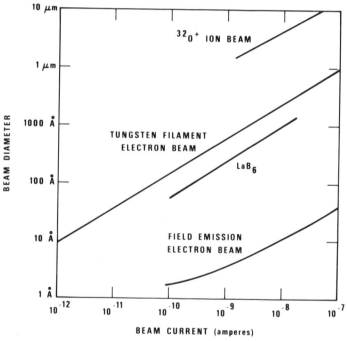

Fig. 3. Analytical beam diameter as a function of beam current for electron and ion beams.

40

A. Electron-Beam Excitation

When an electron beam of 1 to 100 keV interacts with a solid, both elastic and inelastic events occur. This is shown in Fig. 4, where it can be seen that as a result of this excitation there are a large number of emissions that result from these interactions. The most important ones, as far as chemical analysis is concerned, are the x rays, Auger electrons, and those transmitted electrons that have lost energy through inelastic interaction with inner shell electrons of target nuclei, i.e. energy loss electrons. As can be seen in Fig. 5 there are significant differences in volume of excitation among analytical techniques and this will be reflected in differences in elemental detection limits.

Another dominating factor in detection limits for electron-beam excitation is the source of electrons. As shown in Fig. 3 and Table I there are significant differences in beam current and, therefore, number of electrons for excitation of the sample, and these depend on the type of electron source. At the higher beam densities there is also considerable sample damage. This is particularly true for the field emission sources where sample heating can be extreme. Fortunately, single-crystal silicon is very tolerant to high beam currents but caution must be exercised when analyzing silicon surfaces for other materials, e.g., organics.

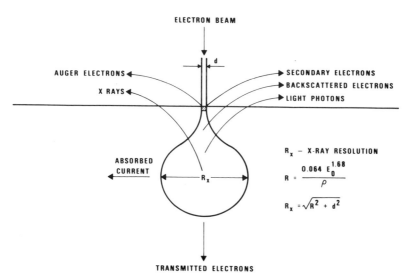

Fig. 4. Illustration of the interaction of an electron beam with an infinitely thick solid, illustrating volume of interaction and species emitted from the excited volume.

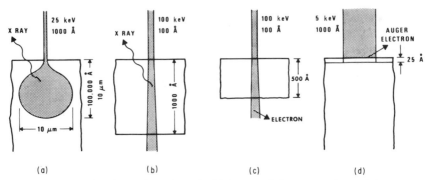

Fig. 5. Illustration of the effect of sample thickness and characterization technique on resultant analytical diameter for interaction of electrons with solids: (a) electron microprobe; (b) scanning transmission electron microscopy; (c) electron energy loss; (d) Auger electron.

TABLE I

Differences in Electron Excitation Sources

Electron source	Current density (A/cm²)	Brightness (A/cm² sr)
Tungsten filament	5–25	1×10^6
LaB$_6$	25–100	5×10^6
Field emission	10^5–10^6	1×10^8

1. Diameter of Analysis Region

When x rays are detected from a thick sample, the volume excited is much larger than the diameter d of the excitation beam. The diameter of the excited volume R_x is shown in Fig. 4 and, when expressed as microns, is a function of the excitation energy E_0 (in kiloelectron volts) and the density of the sample ρ (in grams per cubic centimeter) [4]. Typically for silicon this can range from 0.4 μm for 5-keV excitation up to 20 μm for 50 keV for beam diameters of $\sim$ 2000 Å or less. When x rays are detected and the sample is infinitely thick, the technique is referred to as electron microprobe analysis (EMP).

If the sample is made thin, i.e., $\sim$ 1000–2000 Å, then the diameter of the excited volume emitting x rays is much closer to the diameter of the electron excitation beam. This is shown schematically in Fig. 5 and labeled scanning transmission electron microscopy (STEM). Thin specimens are always used in transmission electron microscopy and generally $\sim$ 100-keV electron beams are used for excitation. As a result the electrons can exit the other side of the specimen before significant spread due to elastic scattering can occur. There can be beam spread [5], where the amount of spread b is a function of excitation energy E (in electron volts), sample density ρ, beam diameter d, atomic number Z, atomic weight A, and sample thickness, such that

$$b = 6.25 \times 10^5 \frac{Z}{E} \left(\frac{\rho}{A}\right)^{1/2} t^{3/2}. \tag{1}$$

For very thin specimens, 100–500 Å, the transmitted electrons will contain electrons that have lost energy due to inelastic collision with bound electrons of nuclei in the sample. This technique is termed electron energy loss spectroscopy. The diameter of the analyzed volume will be close to the electron excitation beam diameter as shown in Fig. 5. This ensures high spatial resolution for this technique.

When an atom that was excited due to loss of a core electron returns to the ground state, two competing processes are involved. This is shown schematically in Fig. 6. Emission of x rays occurs when an outer electron drops from a higher level and the energy of the x ray is equal to the difference in energy between the two electron levels. However, an energy transfer can occur between two other electrons, as shown in Fig. 6. The electron giving up the energy then drops to fill the vacancy. The electron receiving the energy is ejected from the atom with an energy equal to the difference in energy between the electron levels minus the energy necessary to escape the atom. These latter electrons are referred to as Auger electrons, after the French physicist Pierre Auger, and generally range in energy from 20 to 2000 eV. The mean path of Auger electrons is 10–30 Å.

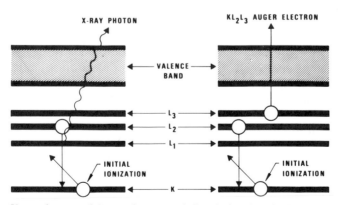

Fig. 6. X-ray photon and Auger electron emission during deexcitation of an atom after initial ionization.

TABLE II

Analytical Diameter Capabilities of Electron-Beam
Characterization Techniques

Technique	Range	Typical diameter
Electron microprobe	1–25 μm	5 μm
Auger spectroscopy	25–300 μm	100 μm
Scanning Auger microprobe	500–3000 Å	2000 Å
Scanning transmission microscopy	50–2000 Å	1000 Å
Electron energy loss spectroscopy	10–100 Å	50 Å

As a result, Auger electrons are highly localized around the diameter of excitation. Thus the diameter of the analysis region in Auger spectroscopy is close to the diameter of the electron excitation beam. In conventional Auger spectroscopy the beam diameter is large, ≥ 25 μm, while in scanning Auger microprobes the beam diameters range from 500 to 3000 Å. Spatial resolutions as small as 300 Å are attainable [6].

The x-ray and Auger deexcitation processes are competitive, with Auger electron emission dominant in the low-Z elements and x-ray emission dominant in the high-Z elements. There is about equal probability for both processes around arsenic ($Z = 33$) in the periodic table.

A summary of the analytical diameter capabilities with typical ranges and typical values is given in Table II.

2. Depth of Analysis Region

In chemical analysis for VLSI it is important to obtain in-depth compositional information. Of the electron-beam techniques, only Auger spectroscopy has the small depth resolution required and this is because of the small range of Auger electrons. This technique, when combined with inert gas ion sputtering, can provide in-depth information as deep as ~ 5000 Å with ~ 25-Å depth resolution. Depth resolution tends to degrade somewhat at depths greater than 1000 Å. The other electron-beam techniques cannot provide this type of information even when combined with an ion sputtering capability.

3. Detection Limits

The major challenge associated with submicron characterization is achieving elemental detection limits that are useful to support VLSI device, materials, and process research and development. As discussed earlier (Section II.A), the best detection limits are realized when the maximum number of electrons are exciting the sample. This quite often means analytical diameters substantially larger than 1 μm. Typical detection limits for electron-beam techniques are shown in Fig. 7 as a function of the analytical diameter. The two lines on the left of this figure show the concentration and analytical diameter where only one impurity atom is present for depths of 25 and 100 Å. It is unlikely that devices can be built with these concentrations and dimensions because of the statistical uncertainty of having one atom present in that specific device element. It is immediately apparent that the detection limits for electron-beam techniques are not useful for the normal dopants or trace impurities associated with semiconductor grade silicon. However, the absolute sensitivity of many of these techniques is outstanding, particularly in microvolume analysis, and this is why they find wide applicability in specific areas. Table III lists the detection limit range and absolute sensitivity, in atoms, for each electron-beam technique.

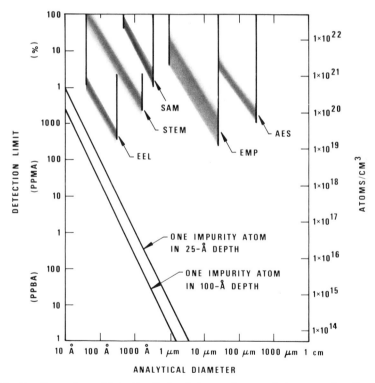

Fig. 7. Comparison of detection limits of electron-beam techniques as a function of analytical diameter. Notice the two lines at lower left defining concentration and diameter where only one impurity is present in analytical depths of 25 and 100 Å.

TABLE III

Detection Limits and Absolute Sensitivity for
Electron-Beam Characterization Techniques

Technique	Detection limit range (%)	Absolute sensitivity (atoms)
Electron microprobe	0.1–10	10^9–10^{11}
Auger spectroscopy	0.5–10	$\sim 1 \times 10^{10}$
Scanning Auger microprobe	1–50	$\sim 1 \times 10^5$
Scanning transmission microscopy	0.5–100	10^3–10^6
Electron energy loss spectroscopy	0.1–5	$\sim 4 \times 10^2$

4. Relevance to VLSI

This high absolute sensitivity is particularly useful in the analysis of decorated dislocations, microprecipitates, epitaxial stacking faults, and other microdefects in processed slices and finished devices. In these cases the total number of atoms present is small, i.e., 10^2–10^6, however, in terms of actual concentration, it is usually found to be close to the pure element, e.g., ~ 50–100%. Similarly, in device surface analysis, it is necessary to analyze only the outer 50–200 Å to understand why a wire bond to a metal bond pad is failing, or an ohmic contact into an etched-out oxide window is rectifying, or what is causing corrosion of aluminum metallization on one portion of a circuit, or why a platinum Schottky diode is failing. In all these cases 5 to 20 monolayers of essentially pure carbon, oxide, nitride, etc., are present but only 10^5 to 10^7 atoms are in the analytical region for detection. Thus these electron-beam characterization techniques are widely used in support of today's research and development in the electronics industry. They obviously will find increasing utilization in the era of submicron VLSI devices.

B. Ion-Beam Excitation

The mechanism of interaction of ion beams with solid surfaces is substantially different than with electrons and the resultant means of materials characterization tend to be quite different. Generally, electron beams interact with the electrons of the sample nuclei and, because of their small mass, do not cause sputtering or ejection of nuclei from the sample. On the other hand, ion beams can range in mass from hydrogen ($Z = 1$) through cesium ($Z = 55$) with energies ranging from 1 keV to 4 MeV (see Table IV). Ion sputtering prevails in the lower energy ranges, i.e., ~ 1–30 keV, and when the positively or negatively charged secondary ions are analyzed using a mass spectrometer, the technique is referred to as secondary ion mass spectroscopy (SIMS). When the exciting ion beam is kept small, i.e., 2.5–10 μm in diameter, and rastered over the sample, the technique is frequently referred to as ion microprobe mass analysis (IMMA).

In the ion-beam energy range of 30 to 300 keV, there is substantially less ion sputtering and the predominant ion–solid interactions result in ion implantation. This energy region is generally not used for the characterization of materials. At the higher energy ranges, 300 keV to 4 MeV, the energy range of the ion in the solid becomes substantial, $\geq 0.1-1$ μm. In this range ion–nucleus interactions can result in Rutherford backscattering of the exciting ion where the energy of the backscattered ion is related to the mass of the target nucleus through Eq. (2):

$$E = [(M - m)/(M + m)]^2 E_0, \tag{2}$$

where M is the mass of the target atom, m the mass of the ion backscattered at 180°, and E_0 the initial energy of the incident ion. It is apparent from Eq. (2) that the exciting beam energy must be as nearly monoenergetic as possible to ensure accurate determination of the target mass. As the monoenergetic beam enters the sample, it will immediately start losing energy through interactions with atomic electrons until it undergoes a Rutherford collision or comes to rest in the sample. The backscattered ion will lose energy in the same way on the way out of the sample. As a result the energy of the emitted backscattered electron will be E_0 less the energy lost entering, less the energy lost in backscattering, less the energy lost exiting. For any given matrix, e.g., silicon, the energies lost entering and exiting per unit length are well known and thus the distance traversed by the ion before and after collision is readily calculated. Therefore, from an ion backscatter energy spectrum, it is possible to obtain the mass and number of target atoms causing backscatter and the depth within the

TABLE IV

Applications of Ion-Beam Interaction with Solids as a
Function of Energy and Ion Type

Energy range	Typical ions	Application
0.5–5 keV	Ar^+, Kr^+, Xe^+	Ion sputtering for profiling
1–30 keV	O_2^+, Cs^+, Ar^+, O^-	Secondary ion mass spectroscopy
30–300 keV	Any	Ion implantation
300 keV–4 MeV	He^+, (H^+, C^+, N^+)	Rutherford backscattering spectroscopy
1.5 MeV and up	H^+, Li^+, C, F	Proton x-ray fluorescence and nuclear reactions

sample where backscattering occurred. Depth resolution is controlled by the energy resolution of the detection system used to measure the energy of the backscattered ion and, as normally practiced, in-depth resolutions of ~ 100 Å are obtained. There are techniques to improve this to ~ 30 Å. Sensitivity is proportional to $Z_0 Z_m / E_0$, where E_0 and Z_0 are the energy and mass of the incident ion, respectively, and Z_m the mass of the target atom. Generally 1–4 MeV helium ions are used in most backscattering characterization work.

1. Diameter of Analysis Region

As shown earlier in Figs. 2 and 3 the diameters of ion beams used in materials characterization are not submicron. The smallest diameters employed in secondary ion mass spectroscopy are ~ 2.5 to 10 μm and for best elemental sensitivities are in the 100-μm-diameter range. In Rutherford backscattering spectroscopy, helium-ion-beam diameters of 100 μm to 1 mm are generally employed. For VLSI materials characterization of materials and devices, there is an obvious need for improvement. Seliger *et al.* [7] have reported Ga^+ ion beams as small as 1000 Å in diameter with beam currents of 1.5 A/cm². These have not been employed for materials characterization but do demonstrate submicron feasibility. It is not clear that Ga^+ ion excitation will produce sufficient positive or negative secondary ions to be useful as a characterization tool elemental analysis for VLSI.

2. Depth of Analysis Region

When VLSI device structures are scaled to submicron dimensions, the depths of ion implantations and thicknesses of dielectric layers will also decrease. More important, the depth distribution of impurities in these more shallow structures must be understood. As a result, in-depth resolutions of 20 to 100 Å with ranges of 100 to 2000 Å will be required.

In secondary ion mass spectroscopy, over 90% of the secondary ions are emitted from the outer two atomic layers of the sample. This means that the sputter rate and time of sampling during in-depth profiling will be the dominant factors in depth resolution. Sputter rates of 2 to 5 Å/sec with data acquisition times of 10 sec means that typical depth increments are in the 20–50-Å range. Ranges of up to 10,000 Å are possible with this technique. However, depth resolution tends to degrade with the deeper analysis because of ion mixing, atom knock-on, etc.

Rutherford backscattering in-depth resolution is generally ~ 100 Å for an incident beam because of the inherent resolution of current silicon detectors. This resolution can be improved to ~ 30 Å by using low-angle bombardment. The range of the helium ions is strongly dependent on the matrix but for low-Z elements such as silicon useful information can be obtained from as deep as 10,000 Å. In-depth characterization is one of the major strengths and areas of application of ion-beam techniques.

3. Detection Limits

Secondary ion mass spectroscopy is generally regarded as a high-sensitivity characterization technique with in-depth profiling capability. In order to achieve this high sensitivity, it is necessary to analyze large areas, as shown in Fig. 8. However, this is the only characterization technique with the capability of detecting dopant levels, i.e., $< 1 \times 10^{18}$ atoms/cm^3 and is therefore widely used for this purpose. One of the major problems associated with the application of this technique is secondary ion mass interference, e.g., $^{30}Si^1H^+$ versus $^{31}P^+$, $^{30}Si^{29}Si^{16}O^+$ versus $^{75}As^+$. There have been three direct attacks on the problem of secondary

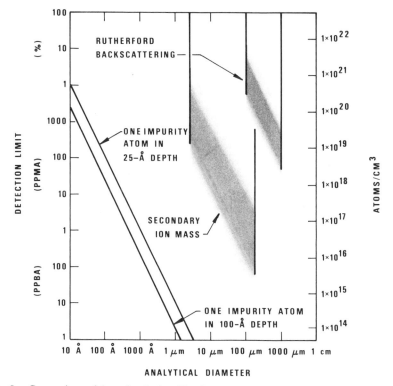

Fig. 8. Comparison of detection limits of ion-beam techniques as a function of analytical diameter.

ion interference (i) higher vacuums—10^{-9}–10^{-10} Torr, (ii) higher resolutions—$M/\Delta M$ 3000–5000, and (iii) alternate ion sources—Cs^+; all of which have helped alleviate the problem. The use of the Cs^+ ion source has been particularly successful because it generates good yields of negative secondary ions and it does not contribute to the vacuum in the same way as gas sources from duoplasmatrons, e.g., $^{32}O^+$.

Typical detection limits for Rutherford backscattering spectroscopy are also shown in Fig. 8. This technique finds wide application in the measurement of ion implantation profiles where peak concentrations lie in the 10^{20}–10^{22} atoms/cm^3 range. The minimum detection limit for this technique is in the 5×10^{18} atoms/cm^3 range.

4. Relevance to VLSI

As discussed earlier, in-depth distributions of dopants in semiconductors and dielectric films are becoming increasingly important in VLSI. Ion implantation will displace diffusion technology and it will be important to understand and follow the dopant distribution after implantation and during device processing. Only the two ion-beam techniques discussed in this section have this capability. Electron-beam excitation techniques, even when coupled with ion sputtering, lack the elemental sensitivity. As can be seen, these ion-beam techniques support process research and development rather than materials or devices. As a result the typical ≥ 100-μm x–y spatial resolution associated with these techniques is not a limiting factor. Process development is done at the slice level. Nevertheless, application to current generation devices is possible utilizing the larger structures on the device, e.g., bond pads or capacitors. In the long run, for submicron VLSI devices, it will be necessary to have ion sources with submicron beams and high beam currents.

C. X-Ray Excitation

The use of x-ray excitation for characterization of VLSI devices, materials, and processes would at first appear to be of little value because x rays cannot be focused and thus all techniques would have large associated analytical diameters (see Fig. 2). Because of the surface analysis (~ 25 Å) capability of x-ray photoelectron spectroscopy and its ability to establish chemical bonding information, there are specific applications that make it unique for VLSI applications.

When a sample is excited by x radiation, the atoms emit photoelectrons in the energy range 10–1400 eV. If the excitation is done with monoenergetic x rays, then the kinetic energy of the emitted photoelectron will be equal to the x-ray excitation energy $h\nu$ minus the binding energy E_b of the electron minus the work function Φ of the spectrometer. Since both Φ and $h\nu$ are known, the binding energy for the electron from a particular shell of the atom can be determined. That binding energy is influenced by the type of bonding or chemical compounds by which the atom is involved. For example, it is possible to differentiate among elemental silicon, silicates, organo silicates, silicon halides, etc., because of the capability of accurately measuring the binding energy.

In the same way, as was described in Section II.A, after photoelectron emission, the excited atom can return to the ground state via either Auger electron or x-ray photon emission. Generally, Auger spectroscopy is not practiced using x-ray excitation; however, Auger electrons are detected in the spectroscopic analysis of the photoelectrons. A better signal-to-noise ratio is obtained for these Auger electrons because of the much lower electron density (noise) in x-ray excitation over electron excitation. However, the total yield of Auger electrons (signal) is smaller than with electron excitation.

When x-ray spectroscopy is carried out on the excited sample, the technique is referred to as x-ray fluorescence. Emitted x rays are energy analyzed using either solid-state Si(Li) nondispersive detectors or energy-dispersive crystal spectrometers. Typical wavelength-dispersive spectrometers have energy resolutions of 1 to 10 eV while the Si(Li) detectors have at best 145–150-eV resolution. The advantage of the solid-state detector is a factor of $100\times$ faster data acquisition. However, the disadvantage is frequent spectral interferences due to the poorer energy resolution.

1. Diameter and Depth of Analysis Region

Since x rays cannot be focused, the only way to examine small areas is by collimation of the excitation beam. In most work the smallest beam diameter used is ~ 1 mm but analyses are generally carried out with $\gtrsim 1$-cm beams. Work with synchrotron radiation offers the best opportunities for more intense, smaller-diameter energy beams [8]. It also is possible to choose the exact excitation energy to optimize analytical sensitivity. A laser x ray would truly revolutionize x-ray excitation techniques.

Photoelectrons have essentially the same range as Auger electrons and, as discussed in Section II.A.2, the depth of analysis is controlled by this range. Photoelectron spectroscopy is obviously a surface analysis technique with an in-depth resolution of 10 to 30 Å, depending on the energy of the emitted photoelectron. When combined with ion sputtering, in-depth ranges of ~ 0.5 to 1 μm are possible.

The fluorescence of x rays exploits the ability to excite and emit to the depth range of the x rays involved. This is, of course, a function of the density of the matrix under investigation. For silicon, depths of 10 to 30 μm are analyzed.

2. Detection Limits

Spectroscopy of x-ray-excited photoelectrons has essentially the same detection limits as conventional Auger spectroscopy, i.e., ~0.5–1%. The effect of the diameter of the analytical area is shown in Fig. 9. Also shown in this figure are the detection limits for x-ray fluorescence and, as can be seen, it is possible to measure concentrations in the low ppma range. The primary reason for this is the deeper analytical region, 10 μm versus 10 Å for photoelectron spectroscopy.

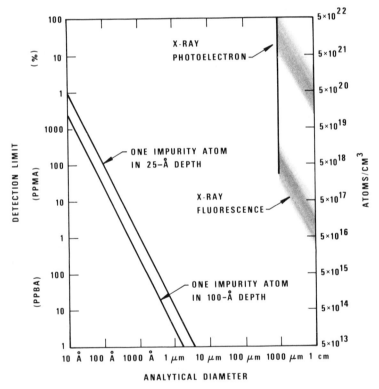

Fig. 9. Comparison of detection limits of x-ray excitation techniques as a function of analytical diameter.

3. Relevance to VLSI

The two x-ray-excited techniques under consideration have vastly different applications when applied to VLSI research and development support. The x-ray-excited photoelectron spectroscopy is a surface analytical tool but finds different applications than the Auger spectroscopy for two reasons. The inability to analyze small areas precludes its use on device structures where microbeam Auger spectroscopy finds wide application. It is, therefore, more widely used in process development where large areas are available for analysis. Second, the ability to identify specific chemical species as opposed to elemental identification makes it ideal for VLSI process research and development and process control. Frequently it is not enough to know that there is a problem due to carbon contamination. It is essential to understand if it is a hydrocarbon, metal carbide, residual photoresist, elemental carbon, etc. Photoelectron spectroscopy can provide this information and can do it as a function of depth into the specimen.

The area of applicability of x-ray fluorescence is also in VLSI process research and development and process control. It is not a surface technique but is ideally suited to thin-film analysis because of its inherent higher sensitivity. The analysis of evaporated aluminum films doped with copper or silicon to control electromigration is a typical example. Compositional analysis of Ti–W sputtered films and platinum and molybdenum thicknesses prior to silicide formation are other applications that do not need high spatial or depth resolution.

III. IMAGING

A. Surface Imaging

Imaging for VLSI research and development is an exceedingly challenging area, as shown in Fig. 1 and as discussed in Section I. The resolutions required in VLSI are far beyond the capabilities of optical microscopy. As shown in Table V, the best resolution attainable using optical microscopy is of the order of 2000 Å. Secondary electron microscopy (SEM) has resolutions 100 times better, which makes it ideally suited to the imaging of surfaces, particularly device features, at small dimensions. As a result of this capability SEM instruments have virtually replaced optical microscopes and resolutions of 30 to 50 Å are routinely obtained in today's instruments. Resolution ~ 10 Å have been obtained on top of the

TABLE V

Comparison of Imaging Resolution as a Function of
Technique and Detected Species

Technique	Detected species	Best resolutions
Optical microscopy	Light	2000 Å
Secondary electron microscopy	Secondary electrons	10–30 Å
Transmission electron microscopy	X rays	500–1000 Å
Electron energy loss microscopy	Transmitted electrons	50–100 Å
Scanning Auger microscopy	Auger electrons	300 Å
Electron microprobe	X rays	0.2–2 μm
Secondary ion mass spectroscopy	Secondary ions	1–5 μm

line instruments using field emission electron sources. Scanning electron microscopy as practiced today can meet the needs of VLSI over the next decade.

B. Chemical Imaging

In addition to physical imaging of a surface, it is frequently necessary to know the distribution of a chemical species in the x–y plane. This can be accomplished using electron- and ion-beam excitation with detected species and resolutions as shown in Table V. The ultimate resolutions of the electron-beam excitation techniques are controlled by the detected species, as discussed in Section II. The best chemical imaging is obtained when thin specimens are analyzed, e.g., electron energy loss and scanning transmission electron microscopy. Excellent chemical imaging of shallow depths on surfaces is attainable in scanning Auger microscopy because small-diameter beams are used for excitation and the Auger electrons are only emitted from the outer 10–30 Å of the specimen.

Ion-excited chemical imaging is controlled by the ion optics in ion microscopy instruments and by the ion-beam diameter in ion microprobes. While ion techniques have the best elemental sensitivites for imaging, the best attainable resolutions are in the 1–5-μm range. There is obvious need for research and development in this area.

IV. PHYSICAL DEFECT CHARACTERIZATION

The detection, imaging, chemical analysis, and electrical activity of physical defects will become of paramount importance in VLSI materials and devices. Physical defects range in size from ~ 10 Å for point defects up to micron-sized stacking faults and precipitates. These defects have played a role in device yields, performance, and reliability in LSI technology where design rules have ranged from 5 to 10 μm. As design rules for VLSI shrink toward micron and submicron dimensions, the minimum feature sizes of devices will be the same or smaller than those of the physical defects. Many of these defects are electrically active and some act as electron generation sites. These types of defects are particularly deleterious to CCD and MOS devices. At present CCD charge wells contain only 100K electrons and are projected to contain less than 50K electrons in future devices. Similarly, MOS memories store ~ 1M to 5M electrons and will store smaller numbers in future generations. The vulnerability of these devices to alpha particles [9], which produce soft errors from the hole–electron pairs produced when the alpha particles interact with the device, bear witness to the detrimental effects of small electrically active physical defects. Complete characterization of these defects requires the application of a number of complementary characterization techniques.

A. X-Ray Topography

Both transmission and reflection x-ray topography are used to obtain large area topographs of physical defects. Figure 10 shows schematically how a single-crystal slice is aligned on a specific Bragg angle such that a collimated, monochromatic beam of x rays will transmit the specimen. There will be anomalous diffraction when the x-ray beam interacts with dislocations in the crystal and these will be recorded on the photographic emulsion. By oscillating the crystal and film together, the entire specimen

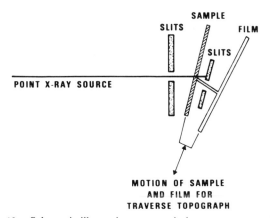

Fig. 10.   Schematic illustrating a transmission x-ray topography.

can be examined. In analyzing large-diameter slices, e.g., 100 mm, which have undergone thermal processing, it is necessary to continuously adjust the specimen to ensure that it stays on the chosen Bragg angle. This can only be accomplished by using computer control. Transmission x-ray topographs provide information on dislocations through the bulk of the crystal. Reflection topographs provide information on the outer 10–30 μm of a crystal surface.

The resolution of x-ray topography is on the order of 1 to 2 μm and this is controlled by a number of factors. The image is formed as a result of diffraction from an anomoly, e.g., strain in the crystal, and as such this technique does not directly image dislocations, stacking faults, etc. Transmission electron microscopy (see Section IV.B) directly images the individual defects.

Figure 11 is a typical example of a transmission x-ray topograph of a 3-in. silicon slice used for the manufacture of 16K DRAM devices. Before topography all metallization, oxide, and nitride had been removed. The strain from the ion implantations and diffusions makes it possible to clearly image the individual devices. The arrays of dislocations on slip planes at the periphery of the slice are a result of stress introduced through improper thermal processing, i.e., too rapid heating or cooling.

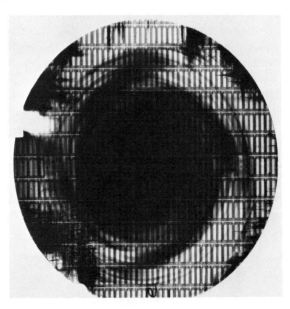

Fig. 11. Transmission x-ray topograph of a silicon slice used for the manufacture of 16K DRAM devices.

The precipitation of oxygen in the Czochralski silicon slice is shown in the center of the slice.

The size of the individual silver grains in the photographic emulsions are on the same order ~ 1 μm and this also limits resolution. The detectability or sensitivity to defects is controlled by the signal (defect)-to-noise (fluctuations in Bragg angle deflected) ratio and improvements are only possible by using double-crystal x-ray topography. This is experimentally an extremely difficult technique and not widely employed. For VLSI defect studies it will be necessary to develop sophisticated computer-controlled double-crystal x-ray techniques.

B. Transmission Electron Microscopy

In order to fully understand physical defects and their role in VLSI materials, processes, and devices, it is essential to obtain high-quality images with resolutions in the tens of angstroms. This can only be accomplished using transmission electron microscopy. Recent extensions of transmission electron microscopy have created instruments with the required image resolutions and a complete analytical capability [10]. The scanning capability using an intense electron beam from a field emission gun makes possible transmission image resolutions of 2 to 5 Å, secondary electron image resolutions of ~ 10 Å, x-ray analysis of spots as small as 25–30 Å, electron energy loss analysis for light elements in spots of ~ 50 Å, and microbeam electron diffraction of defects of ~ 30 Å. These capabilities are extremely important for VLSI defect understanding.

1. Defect Imaging

Physical defects, their formation and role in VLSI devices, are not well understood in the electronic materials under consideration for VLSI, e.g., Czochralski silicon, float-zone silicon, silicon on sapphire, and gallium arsenide. The defects are not present in the starting materials but are abundant after processing. They are formed when the lattice is strained during thermal processing for anneals and oxidations. They are also introduced through ion implantation and diffusion strain, from polysilicon and thick oxide patterning and in situ precipitation of species such as oxygen in the bulk. The real challenge is to isolate, image, and understand the formation mechanism and role of the physical defect in the materials and final devices. These physical defects can take the form of microprecipitates, inclusions, stacking faults, edge and screw dislocations, decorated and pure dislocations, Frank and Shockley partial dislocations, interstitials, and vacancies. Physical defects can be a combination of multiple physical and chemical defects and high-resolution imaging is the only way to understand their behavior.

An example of such a complex defect is shown in Fig. 12. It was known that this defect was an electron generation site and as such was particularly deleterious to MOS DRAMs and CCD devices. From the image it can be deduced that the defect was initially a stacking fault that was partially unfaulted, leaving a pure dislocation and a decorated dislocation. The size of the precipitates on the decorated dislocation range in size from 50 to 100 Å in diameter. Placing the electron beam in the center of one of these precipitates and monitoring the x-ray emission clearly showed the precipitates to be clusters of 5000 to 50,000 atoms of copper. This explains the electrical activity of the defect as an electron generation site. The source of the copper was subsequently traced to a contaminated quartz furnace tube.

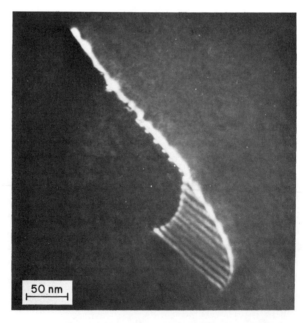

Fig. 12. Electron micrograph of an electrically active physical defect in silicon showing a copper precipitation decorating a dislocation on an unfaulted stacking fault.

2. Chemical Analysis of Defects

From the preceding discussion it can be seen that chemical analysis along with high-quality imaging is necessary to understand the behavior of the defect. In general the physical size and quantity of impurities associated with defects are extremely small. Only high-resolution electron-beam techniques associated with the imaging process can perform this demanding characterization. This is restricted to scanning transmission electron microscopy where the high-intensity electron beam can be focused on specific features of the physical defect. When emitted x rays are detected, a Si(Li) detector with a beryllium window will detect x rays from elements with atomic number Z higher than sodium ($Z = 11$) in the periodic table. Most decorated physical defects in silicon do not emit detectable x rays when excited. This means they have atomic numbers less than sodium and are probably carbon or oxygen. An obvious need exists for the use of windowless Si(Li) detectors in these instruments.

An electron energy loss spectrometer is an obvious addition to scanning transmission electron microscopes. The inherent sensitivity of electron energy loss spectroscopy to the low-Z elements makes it a complementary tool to x-ray detectors. The problem has turned out to be that extremely thin samples must be utilized, e.g., $\sim 100-500$ Å. The preparation of these thin specimens in a manner that will preserve the physical defect under investigation has generally not met with a great deal of success. As better sample preparation techniques are developed, electron energy loss spectroscopy will receive wider application. There still remains considerable work to be done on the technique through computer processing of both spectra and images.

C. Ion Backscattering

It is frequently necessary to establish the distribution of physical defects ranging from the surface into the bulk of the crystal. This type of information is necessary in order to measure the effectiveness of, for example, thermal or laser annealing after ion implantation, laser annealing for epitaxial silicon growth, or annealing of silicon on sapphire. When a silicon crystal is aligned on a major crystalline axis, the $2-4$-MeV He$^+$ ions used in Rutherford backscattering will penetrate deeply into the crystal with very little backscatter. However, if there are silicon atoms or impurity atoms in interstitial positions, He$^+$ ions will be backscattered. Using this technique it is possible to tell what interstitial atoms are causing the backscatter and at what depth they are in the specimen. When this measurement is made before and after thermal annealing, it is possible to obtain in-depth information on disorder and the effectiveness of the anneals. This technique has been widely used in the transformation of polysilicon layers to single-crystal films using laser annealing [11].

D. Electron-Beam Excitation

The establishment of the electrical activity of physical defects is vital to the understanding of their role in device performance and reliability. Electron-beam probing, or excitation, is an ideal tool because of its high spatial resolution capability with small beam diameter and the ability of the high-energy electrons to excite the defect. After excitation the radiative or nonradiative recombination properties and energy level of the defect can frequently be determined.

While the exciting electron-beam diameter can be small, i.e., < 1000 Å, the volume excited is large and spatial resolution is in the micrometer range, as was discussed in Section II.A and shown in Figs. 4 and 5 for the electron microprobe. As described earlier, if the sample is very thin, e.g., as used in transmission electron microscopy, then there is less beam spreading and better spatial resolution is obtained. Resolution can also be improved by working at lower beam voltages, e.g., 5–10 keV for thick specimens or at high beam voltages 50–200 keV for thin specimens.

1. Electron-Beam-Induced Current

The technique of electron-beam induced current requires the presence of a p–n junction, a Schottky diode, or an MOS capacitor in order to collect the hole–electron pairs generated by the exciting electron beam [12]. If there are no perturbations that disturb the electron–hole balance, then the signal that is collected and displayed on a CRT screen and synchronized with sweeping the electron beam will not show any contrast. If there is localized recombination through localized defects, diffusion-induced dislocations, or traps in the space-charge region, they will be imaged. While these images are qualitative in nature, they are readily correlated with specific device and/or material defects. This technique provides a valuable tool for evaluating materials, processes, and devices for VLSI. It will receive even more application to support research and development as device design rules shrink and electrically active defects play a more dominant role than in devices designed with today's 5-μm design rules.

2. Scanning Deep-Level Transient Spectroscopy

Scanning deep-level, transient spectroscopy is an extention of the technique to be described in Section V.A. In this case the energies of the deep levels are determined in concert with the scanning electron beam in a transmission electron microscopy [13]. The deep-level transient signal is displayed on a CRT screen as the electron beam is scanned across the sample and simultaneously the temperature of the specimen is varied from 40 to 600 K. In this way images at specific temperatures correspond to high-resolution maps of specific deep levels. The spatial resolution was optimized by using high-energy electron beams (50–200 keV) to minimize electron-beam spreading in the chemically thinned epitaxial layers under investigation. The ultimate resolution for this technique is the same as for electron-beam-induced current (discussed in the preceding subsection) and cathodoluminescence (discussed in the following subsection) and is on the order of the diffusion length or the excited volume. Generally this is in the $0.1–2$-μm range.

The sensitivity of this technique is strongly dependent on the excited volume and, with a resolution of 20 μm, is $\sim 10^{15}/cm^3$. This needs to be improved for VLSI quality devices and materials, e.g., silicon.

3. Cathodoluminescence

The III–V and II–VI semiconductors emit light during radiative recombination of the holes and electrons generated by the electron beam [14]. When the signal from the optical detector is displayed on a CRT screen synchronized with the sweep of the electron beam, an image of the recombination center is obtained. As in other electron-beam excitation techniques, the volume excited is small and adjusting the signal-to-noise ratio is a constant problem. Resolutions of 0.5 to 5 μm are obtained on these types of semiconductors.

This technique is not applicable to silicon and will be useful in VLSI research and development as work on the new high-speed GaAs submicron logic and memory devices evolves.

V. ELECTRICAL/FUNCTIONAL PROPERTIES

The combined effects of inadvertent impurities, intentional dopants, and physical defects will be reflected in the yield, performance, and reliability of VLSI devices. These effects will become more and more dominant as VLSI design rules reach 1 μm and smaller. However, extraction of materials information on the impurities or defects from measurement of the functional property, e.g., capacitance, resistivity, and lifetime, at submicron dimensions is not readily achievable. Normally the integrated or system function of the device/material is measured and then correlation with the chemical or physical defects is attempted. This is an indirect approach, and as device complexity increases, these types of correlations become more and more difficult. It appears that for the foreseeable future it will be necessary to incorporate a number of strategically placed and strategically designed test patterns on each slice. These test patterns would incorporate structures of the correct size to maximize the signal-to-noise ratio in the ensuing measurements and to provide insight into problems related to materials, processing, and device reliability. The characterization challenge simplifies to controlling signal to noise. As device geometries are scaled down it will be necessary to optimize signal and minimize noise.

A. Deep-Level Transient Spectroscopy

The measurement of transients in the capacitance and current of the space-charge region depleted of carriers near $p-n$ junctions, Schottky diodes, and MOS capacitors can be used to characterize materials for deep levels [15]. In the case of the diodes, the technique yields the deep-level energy, its concentration, and the cross sections for hole and electron capture. The MOS capacitor structure yields the interface state density and the interface state trap energy and concentration. Deep-level, transient spectroscopy is an extremely sensitive technique with detection limits of 10^{10} to $10^{11}/cm^3$. However, large-area devices are required, i.e., ~ 750 μm in diameter, and when these are scaled to small devices, detection limits become unacceptably poor, particularly for VLSI quality materials and devices.

New higher-sensitivity deep-level spectroscopy techniques are under development. The use of conductance transients in place of capacitance has been reported to be more sensitive [16]. Considerably more work is required on these techniques for an understanding of the role of deep levels in VLSI materials.

B. Lifetime

There is considerable confusion both in the literature and with most scientists who design and build semiconductor devices on what lifetime measurements should be made and on how to relate the measurements to device and materials performance. Lifetime characterization techniques have been developed and are widely used for large bulk crystals, slices, and devices. Unfortunately, there is little correlation on lifetimes measured on these different specimens by the associated techniques. The role of dislocations, stacking faults, oxygen precipitation, and metal impurities tends to be unclear but because they are not understood, they are regarded as deleterious. Clearly, extensive research and development are required in lifetime characterization for VLSI.

VI. VLSI CHARACTERIZATION TECHNOLOGY NEEDS

The characterization technology needs for VLSI materials, processes and devices is clear. Better sensitivities at small geometries for chemical analysis are vital. For submicron characterization, bright, finely focused electron beams will be widely applied. Field emission sources provide the required brightness and new, more efficient detection systems are required. Electron energy loss spectroscopy appears particularly exciting both with regard to spatial resolution and detectability. However, sample preparation techniques and digital spectral processing are needed. It also appears that 300–500-keV electron sources will lower the scattered electron background in electron energy loss spectroscopy and thus enhance the signal-to-noise ratio. The addition of Auger detectors with digital pulse counting to scanning transmission electron microscopes to complement Si(Li) x-ray detectors for the analysis of physical defects must be accomplished in order to observe the light-Z elements. Differentially pumped ion sputtering sources will be necessary to expose the defects.

Ion-beam techniques need to have enhanced spatial resolution, moving from the current 2-μm capability to ~100 Å. As this occurs, brighter ion sources will be required to maximize elemental sensitivities. More efficient detection systems will also be mandatory. Current secondary ion mass spectroscopy instrumentation upgrading trends must continue; e.g., sample vacuum improving to 10^{-10} to 10^{-11} Torr and mass resolution ($M/\Delta M$) increasing from 300 to 10,000 without loss of secondary ion throughput. More computer automation in instrument control, data acquisition, and data reduction are essential to enhance the signal-to-noise ratio. New ion sources, e.g., iodine and gallium, will help alleviate the secondary ion mass interference problems. Ion imaging, utilizing computerized digital image processing, must evolve to allow submicron imaging resolution.

The inability to focus x rays makes it necessary to have extremely bright sources that can be collimated sufficiently to achieve submicron resolutions. Synchrotron radiation from electron storage rings appears to offer this potential; however, these sources will not be available to industrial users for rapid turnaround solutions to VLSI problems. The development of an x-ray laser will revolutionize materials characterization as well as all fields of science including x-ray lithography for VLSI. Higher sensitivity, both in imaging and diffraction, for the detection of physical defects will require the development of computer-controlled, double-crystal, x-ray diffraction systems. VLSI devices will be far more susceptible to microdefect-related failures.

In functional characterization, the same signal-to-noise problem is overriding. As geometries shrink to submicron dimensions, the measurement of lifetime, mobility, deep levels, surface states, surface traps, etc., becomes increasingly difficult. Clearly, there is room for the development of innovative new characterization techniques to measure electrical properties at submicron geometries. Correlation of these measurements with both physical and chemical defects clearly falls in the domain of VLSI materials characterization.

ACKNOWLEDGMENT

The x-ray topograph and transmission electron micrograph were supplied by Dr. Herbert Schaake and his contribution is gratefully acknowledged.

REFERENCES

1. G. B. Larrabee, *Scanning Electron Microsc. 1977*, 639 (1977).
2. P. F. Kane and G. B. Larrabee, "Characterization of Solid Surfaces." Plenum Press, New York, 1974.
3. C. A. Evans and R. J. Blattner, *Ann. Rev. Mater. Sci.* **8**, 181 (1978).
4. D. R. Beaman and J. A. Isasi, "Electron Beam Microanalysis," ASTM STP 506. American Society for Testing and Materials, Philadelphia, Pennsylvania, 1972.
5. J. I. Goldstein and D. B. Williams, *Scanning Electron Microsc. 1977*, 651, (1977).
6. M. M. El Gomati and M. Prutton, *Surf. Sci.* **72**, 485 (1978).
7. R. L. Seliger, J. W. Ward, V. Wang, and R. L. Kubena, *Appl. Phys. Lett.* **34**, 310 (1979).
8. W. E. Spicer, I. Lindau, and C. R. Helms, *Res. Dev.* **28** (12), 20 (1977).
9. T. C. May, *Proc. Electron. Components Conf., 29th* p. 247. IEEE, New York, (1979).
10. D. M. Mayer and D. C. Joy, *J. Met. (U.S.A.)* **29**, 26 (1977).
11. J. C. Bean *et al.*, *Appl. Phys. Lett.* **33**, 227 (1978).

12. H. J. Leamy, L. C. Kimerling, and S. D. Ferris, *Scanning Electron Microsc. 1978,* 717 (1978).
13. P. M. Petroff, D. V. Lang, J. L. Strudel, and R. A. Logan, *Scanning Electron Microsc. 1978,* 325 (1978).
14. L. J. Balk and E. Kubalek, *Scanning Electron Microsc. 1977,* 739 (1977).
15. T. Sah, *Solid-State Electron.* **19,** 975 (1976).
16. L. Forbes and U. Kaempf, *Hewlett-Packard J.* **30,** 29 (1979).

Chapter 3

Physics and Modeling of Submicron Insulated-Gate Field-Effect Transistors. II

Transport in the Quantized Inversion Layer

D. K. FERRY

Department of Electrical Engineering
Colorado State University
Fort Collins, Colorado

K. HESS

Department of Electrical Engineering
and Coordinated Science Laboratory
University of Illinois
Urbana, Illinois

P. VOGL

Institute for Theoretical Physics
University of Graz
Graz, Austria

67

I. INTRODUCTION

A. Small Devices and Two Dimensions

If an electric field is applied normal to the surface of a semiconductor, such as in a metal–oxide–semiconductor (MOS) structure found in the insulated-gate field-effect transistor, the conduction and valence bands are bent and an inversion or accumulation layer is formed. For a p-type semiconductor, for example, either an n-type inversion layer or a p^+ accumulation layer can be formed by biasing the gate either positively or negatively, respectively, with respect to the semiconductor substrate. When the bands are strongly bent, the potential well formed by the insulator–semiconductor surface and the electrostatic potential in the semiconductor can be sufficiently narrow that quantum-mechanical effects in the well become important. The motion of electrons (or holes) trapped within the potential well is constrained, in the direction perpendicular to the surface, to remain within the well and, if the thickness is less than the electron wavelength, size-effect quantization leads to widely spaced subbands, each of which corresponds to a particular quantized level for motion perpendicular to the surface (for a general review of the details of quantization, see Stern [1]). Then the carriers are said to constitute a quasi-two-dimensional electron (or hole) gas confined to a region typically a few tens of angstroms thick at the semiconductor–insulator interface. Once thought to be a characteristic only of very low temperatures, these effects can clearly be observed even at room temperatures, especially in the highly doped submicron devices expected in VLSI. The transport of carriers within the inversion layer is affected by the scattering centers at the semiconductor–insulator interface as well as by the optical and acoustical phonons.

Why should we be concerned with such a pseudo-quantum-mechanical behavior in transport? In fact, first impressions dictate that such effects are probably only important at very low temperatures. However, this is not really the case. For example, let us take the case of a silicon MOS field-effect transistor (MOSFET). In a general study of the fundamental limits of MOS technology in silicon, Hoeneisen and Mead [2] considered the general static limitations of oxide breakdown, source-to-drain punchthrough, etc., on the typical silicon MOS transistor and postulated a minimal device size. In such a device, the source-to-drain spacing was 0.25 μm, the gate oxide thickness was 140 Å, and the substrate doping was $2.7 \times 10^{17}/cm^3$. The actual detailed operation of the device was not considered. However, let us carry the analysis somewhat further. The depletion width is readily found to be about 700 Å, giving a surface depletion charge of about $2 \times 10^{12}/cm^2$. Oxide fields of greater than 2×10^6 V/cm are expected in their analysis [2], so that inversion densities of the order of $10^{13}/cm^2$ are expected; these are greater than the surface depletion charge. Under these conditions, the surface potential must be solved self-consistently, as just indicated, and quantum conditions can be expected. In fact, for a surface field of 2×10^6 V/cm, the excited subbands are found to be *more than* 170 meV above the lowest ground state subband. Hence, the level separation is *more than* $6k_B T$ *at room temperature,* and surface quantization can be expected to be important for device operation. Although this case may seem extreme, it clearly is not. Fang and Fowler [3], in studying quantization and high field transport in silicon MOS devices, readily reached inversion densities of $6 \times 10^{12}/cm^2$ in normal-sized devices at 300 K. In a theoretical discussion of their results, it is clear that over the electric field range investigated the transport is dominated by the subband nature of the inversion layer [4]. Clearly, in extending to the reduced size expected in future devices, quantization can be expected to play an even greater role.

Since we shall need to consider transport in quasi-two-dimensional systems at semiconductor surfaces, we should look into the basic aspects of such transport. First and foremost, this transport is far more sensitive to the nature of the interface and to local charge states near the interface. As such, and because at present we know so little about the nature of the interface itself, this surface transport can even be used as a probe to study the electrical properties of the interfacial region.

B. Nature of Two-Dimensional Quantization

The motion of electrons in the quantized levels in the potential well of an accumulation or inversion layer near the semiconductor surface can be characterized in the effective mass approximation by an envelope function [5]

$$\psi(x, y, z) = \zeta_i(z) \exp(i\mathbf{k} \cdot \mathbf{r}), \qquad (1)$$

where $\mathbf{k}$ and $\mathbf{r}$ are two-dimensional vectors in the plane of the surface and z the direction perpendicular to the surface. The wave functions $\zeta_i(z)$ satisfy the one-dimensional Schrödinger equation

$$\frac{h^2}{2m^3} \frac{d^2\zeta_i}{dz^2} + \{E_i - V(z)\}\zeta_i(z) = 0 \qquad (2)$$

and the energy bands are given by

$$E_i(\mathbf{k}) = E_i + \left(\frac{\hbar^2 k_1{}^2}{2m_1} + \frac{\hbar^2 k_2{}^2}{2m_2}\right) \tag{3}$$

for masses m_1, m_2, and m_3, where 1, 2, and 3 refer to the x, y, and z axes, respectively. Here we have actually simplified Eq. (2) in the following sense. The constant energy ellipsoids of the conduction band are assumed to be oriented such that a principal axis is normal to the surface, so that the coordinate axes (x, y, z) correspond to principal directions of the ellipsoids to avoid cross terms in the Laplacian operator. However, this is the case for a device fabricated on the $\langle 100 \rangle$ surface of silicon. From Eq. (3), it may readily be seen that each eigenvalue E_i is the bottom of a two-dimensional continuum of levels and the set belonging to each E_i is termed a subband.

In actuality, the Schrödinger equation is not complete until the appropriate boundary conditions are specified and the self-consistent potential energy $V(z)$ is determined. The simplest boundary conditions are $\zeta_i(0) = 0$ and $\zeta_i(\infty) = 0$. The first of these assumes an infinite potential barrier at the semiconductor–insulator interface, while the latter assumes that the carriers are, in fact, trapped in the potential well. It is the first of these boundary conditions that constitutes a minor problem, but in silicon the interface is typically very sharp (less than 5 Å). Moreover, there is a step of several electron volts into the SiO_2 bands, so that the assumption of $\zeta_i(0) = 0$ introduces only a minor error. Therefore, the assumption of the first boundary condition can be reasonably made. Where quasi-two-dimensional behavior has been observed, the use of this boundary condition to determine the subband has not generated any large noticeable errors. Why this should be so is not clear, but this question will probably not be resolved until the exact nature of the interface structure is determined.

The potential $V(z)$ is found from Poisson's equation in a self-consistent manner [5]. Considering the uncertainties in the true boundary conditions though, an approximate solution for which an analytic solution is obtainable is usually used. The simplest approach, valid for inversion layers primarily, is to assume that $V(z) = eF_s z$ for $z > 0$ (in the semiconductor) where F_s is the surface electric field and e the electronic charge. This model potential is termed the triangular potential and leads to the Airy equation with the solutions [6]

$$\zeta_i(z) = \mathrm{Ai}\{(2m_3 eF_s/\hbar^2)^{1/3}[z - (E_i/eF_s)]\}, \tag{4}$$

$$E_i \approx (\hbar^2/2m_3)^{1/2}\{\tfrac{3}{2}\pi eF_s(i + \tfrac{3}{4})\}^{2/3}, \tag{5}$$

where m_3 is defined above.

The actual eigenvalues are the zeros of the Airy functions and differ very slightly from the values given in Eq. (5). In Section II.A we actually plot the level structure and widths for a Si inversion layer.

II. REVIEW OF SCATTERING MECHANISMS

The scattering of electrons and holes in bulk silicon has been treated extensively in the past (see, e.g., Jacoboni *et al.* [7]). The main scattering agents are thermal lattice vibrations (phonons) and ionized impurities (point charges). The predominant type of scattering determines the shape of the current–voltage characteristic. For example, if we vary the carrier mean energy, acoustic-phonon scattering (via the deformation potential) leads to sublinear nonohmic behavior, while scattering by point charges leads to superlinear curves, as illustrated in Fig. 1. Scattering by the deformation potential arises due to the fact that the phonons change the distance of neighboring atoms and therefore modify slightly the conduction and valence band energies. In a perturbation theory approach, this modification gives rise to scattering. In polar materials, the electrons are scattered by macroscopic electric fields associated with the polar phonons, which may also lead to a superlinear characteristic. At interfaces or surfaces, these scattering mechanisms have to be modified and additional scattering mechanisms have to be taken into account [8].

The transport of electrons in the inversion layer is a subset of the general problem of surface transport, which has a fairly broad literature [8–11]. Transport effects within the surface channel can be separated from bulk effects by modern integrated circuit techniques [12], and a comprehensive study of transport in inversion layers of silicon was carried out by Fang and Fowler [13]. Theoretical treatments of the scattering pro-

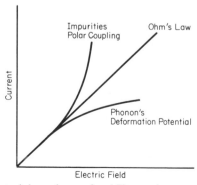

Fig. 1. Conceptual dependence of mobility on the mean carrier energy.

cesses in the inversion layers have dealt with acoustic-phonon scattering by surface [14–16] and bulk [15,16] lattice vibrations, and coulomb scattering near the surface [5,17–22]. At relatively high values of inversion electron densities and over an extended range of temperature, acoustic-phonon scattering and coulomb scattering alone cannot account for the observed mobility.

Scattering due to optical and intervalley phonons was first considered by Sah *et al.* [17] and also by Hess and Sah [23] and Krowne and Holm-Kennedy [24] for the case of hot electrons in the inversion layer. In each case, the phonons involved were treated as bulk phonons and no account of quantization of the final state was taken into account; a reasonable procedure only if all of the electrons are in the lowest subband. Moreover, in the work of Sah *et al.* [17], only a single-optical intervalley phonon was considered and the strength of the interaction was adjusted to provide a fit with data. It is known that the scattering in silicon is more complex than this, and there is no reason that the same coupling constants found in the bulk should not appear in the bulk phonon interaction in the inversion layer. Interactions to both zero order and first order are important [25]. A selection rule, found from the matrix elements of the zero-order interation, prohibits scattering among different subbands belonging to the same set of equivalent valleys in the inversion layer.

In this section, we shall review all the important scattering mechanisms for carriers in the inversion layer. The theory of the individual scattering processes will be discussed and rates determined. The basic quasi-two-dimensional nature of the transport of carriers in the quantized inversion or accumulation layers modifies the dynamics of the various scattering mechanisms. In addition, new scattering processes peculiar to the semiconductor–insulator interface tend to dominate the scattering in strongly quantized levels.

A. Surface Scattering

1. Surface-Roughness Scattering

In a traditional treatment, surface scattering arises from the diffraction of the electron wave function off asperities at the surface (or interface), and the scattering differs in detail for specular or nonspecular scattering [26]. In a quasi-two-dimensional system, however, the motion perpendicular to the surface is constrained. Although several authors have tried to extend the classical picture to a quasi-two-dimensional situation [27–29], the most logical approach follows a random potential argument [30]. Fluctuations in the surface layer (surface roughness) lead to fluctuations in the potential $V(z)$ and in the level width z_i. These in turn cause fluctuations in the various subband energy levels E_i. These potential fluctuations lead to a scattering potential, just as in the case of pseudorandom alloys of III–V compounds [31–33].

We consider that the surface has a random fluctuation $\rho(x, y)$ in the z direction (normal to the surface). Then the potential can be written as

$$V[z - \rho(x, y)] = V(z) + \rho(x, y)F_s + \cdots, \qquad (6)$$

where F_s is the surface electric field. The second term on the right-hand side of Eq. (6) provides the scattering potential and leads to the matrix element

$$\langle \mathbf{k}'|H'|\mathbf{k}\rangle = (e/A) \int \psi^*_{\mathbf{k}'}\psi_{\mathbf{k}}\rho(x, y)F_s \, dA \equiv eF_s\rho_{\mathbf{Q}}, \qquad (7)$$

where $\mathbf{Q} = \mathbf{k}' - \mathbf{k}$, A is the area, and H' the perturbing Hamiltonian. To proceed further, one must make some reasonable assumptions on the nature of $\rho(x, y)$. As in Eq. (7), we retain the perturbing potentials only to first order, and assume that $\rho(x, y)$ is describable by an rms height parameter Δ and a Gaussian autocorrelation in position which is describable by a correlation length L. Then the matrix element becomes

$$|\langle \mathbf{k}'|H'|\mathbf{k}\rangle|^2 = (e^2 F_s^2 \pi \Delta^2 L^2/A) \exp(-Q^2 L^2/4), \qquad (8)$$

where

$$Q = 2k_F \sin(\theta/2) = k_F[2(1 - \cos \theta)]^{1/2}, \qquad (9)$$

θ is the scattering angle, and k_F the Fermi vector. The scattering rate can then be easily found to be

$$1/\tau_{sr} = (m^*\Delta^2 L^2/2\hbar^3)(e^2/\bar{\epsilon})^2(n_s + N_{depl})^2 f(kL), \qquad (10)$$

where Δ and L are the rms height and the correlation length of the roughness, respectively, $\bar{\epsilon} = (\epsilon_{ox} + \epsilon_{sc})/2$, ϵ_{ox} and ϵ_{sc} are the dielectric constants of the oxide and semiconductor, respectively, n_s the inversion density, and N_{depl} the depletion charge. The function $f(kL)$ is a slowly varying function given as

$$f(kL) = \pi \exp(-k^2 L^2/2)\{I_0(k^2 L^2/2) - I_1(k^2 L^2/2)\},$$

where I_0 and I_1 are modified Bessel functions of the first kind. For practical values of kL, $f(kL) \approx \pi$. It may be noted from Eq. (10) that the scattering rate increases rapidly with n_s, since for larger n_s the carriers are pulled closer to the interface. For reference, the values of Δ and L found in Si are typically of the order of 3 and 30 Å, respectively [34].

2. Coulomb Scattering

To first look at coulomb (or impurity) scattering, we examine a single scattering center located at $\mathbf{r} = 0$ and at z_0, where $\mathbf{r}$ is a two-dimensional vector in the x–y plane and the z direction is taken to be the direction perpendicular to the surface. The choice of $\mathbf{r} = 0$ places no restriction on the generality of the problem as long as the scattering centers are distributed homogeneously in the x–y plane. According to Stern and Howard [5], the Poisson equation then takes the form

$$\nabla^2\phi(r, z) - 2 \sum_n S_n g_n(z) \int_0^L \phi(r, z)g_n(z)\, dz = -\frac{e}{\epsilon}\, \delta(r)\, \delta(z - z_0), \quad (11)$$

where S_n is the screening constant (which is given for the first subband by $S_1 = e^2 n_s/2kT\epsilon$ in the two-dimensional limit), $\bar{\epsilon}$ the average permittivity of Si and SiO$_2$, ϵ the permittivity of Si, and $g_n(z)$ the square of the z-dependent part of the wave function. We need to calculate the matrix element of the transition probability that is proportional to the two-dimensional Fourier transform of the potential. Therefore, it is most convenient to solve Eq. (11) immediately by Fourier transformation. We do this by multiplying Eq. (11) by $\exp(iqr)$ and integrating over r. Then one obtains

$$\left[\frac{\partial}{\partial z^2} - q^2\right]\phi(q, z) - 2\sum_n S_n g_n(z) \int_0^L \phi(q, z)g_n(z)\, dz$$

$$= -\frac{e}{\epsilon}\, \delta(z - z_0). \qquad (12)$$

Using the mean value theorem for integrals and the Green's function for the term $[(\partial/\partial z^2) - q^2]$ we have

$$\phi(q, z) = -\int_{-\infty}^{+\infty} dz'\, \frac{\exp(-q|z - z'|)}{2q}$$

$$\times\, [2\Sigma LS_n\phi(q, z_m)g_n(z_m)g_n(z') - \frac{e}{\bar{\epsilon}}\, \delta(z' - x_0)], \qquad (13)$$

where z_m is the suitable mean value of z. Equation (13) can be solved explicitly but the result is lengthy. Usually, $g_n(z)$ is approximated by a δ function and only $n = 1$ is considered. In this way, we obtain a lower limit for the scattering rate (upper limit for the mobility), since the δ-like wave function has the minimum overlap with the potential of the remote scattering centers. With these approximations and a homogeneous concentration N_I of point charges in the x–y plane, one obtains a simple formula for the mobility μ_I [17,35]:

$$\mu_I = (e/m^*)(\bar{\epsilon}/e^2)^2(2\hbar k_B T/\pi^2 N_I). \tag{14}$$

Slightly different formulations have been obtained by Stern and Howard [5] to account for the degeneracy at low temperatures and by Stern [36] to account for detailed screening. Figure 2 shows some experimental data for various densities of point-charge scatterers. Indeed, one can observe $\mu_I \alpha T$ for high densities of scatterers. In view of the complications of Eq. (13) and the possible complicated dependence of N_I on z_m, the good agreement with Eq. (14) seems to be rather accidental and more detailed solutions of Eq. (13) should be used if higher goals are attempted.

B. Phonon Scattering

Numerous papers have been devoted to the scattering of inversion-layer charge carriers by interface and bulk phonons. To obtain a rough understanding, one generally only has to consider two limiting cases. The electrons or holes may form a quasi-two-dimensional system because of size quantization. However, this does not imply that the phonons are in-

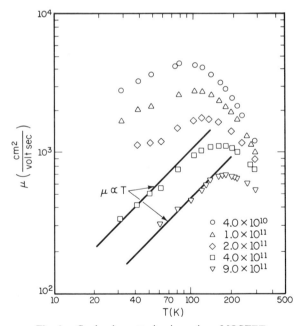

Fig. 2. Coulomb scattering in various MOSFETs.

terface modes. The "two-dimensionality" of the phonons depends on the elastic constants of the two neighboring media, which are only remotely related to the interface potential confining the charge carriers. Therefore, we can treat the charge carriers as two-dimensional, but for the phonons we have to consider: (i) that the elastic (dielectric) constants are close enough and bulk modes exist, and (ii) that the elastic (dielectric) constants are sufficiently different that interface modes exist. This distinction has not always been made in the literature and some of the formulas given are rather meaningless mixtures of these cases. The basic interaction mechanisms (deformation potential and polar coupling) have been applied unchanged to scattering at interfaces, which may not be a valid approximation. The deformation potential concept is based on the assumption that the interaction of electrons with a long wavelength acoustic mode is equivalent to the effect of a locally homogeneous strain. This conjecture (deformation potential theorem) is certainly violated if the width of the inversion layer is of the order of the phonon wavelength, which is the case in wide ranges of temperatures. It is trivial to show that in a square well the vibration of the walls contributes to scattering that is not included in ordinary deformation potential theory [37], but it is not clear how this differs from surface–interface excitations or from surface-roughness scattering. While the formulas that follow represent the best interpretation of current understanding, they should be carefully applied, especially in certain temperature ranges. The polar interaction mechanism has been considered in more detail due to the possibility of scattering by remote polar phonon-generated interface modes [34,38].

1. The Acoustic Interaction

A first principle calculation of the electron–phonon interaction for inversion layers at the $Si-SiO_2$ interface has not yet been performed, and the formulas found in the literature are based on the well-known phenomenological approaches. At the beginning of this section we discussed that the conjecture of the equivalence of phonons and homogeneous strain breaks down when the wavelength is of the order of the inversion-layer width [37]. Over some ranges of temperature, we therefore have to take the deformation potential concept with a grain of salt. However, for small phonon wavevectors, i.e., $|\mathbf{q}| \ll 10^7/cm$, we can write

$$\Delta E = \Xi_1 \nabla \cdot \mathbf{u}, \tag{15}$$

when Ξ_1 is the deformation potential constant, ΔE the change in energy of the subband under consideration, and $\mathbf{u}$ the lattice displacement. For bulk phonons we have

$$\mathbf{u}(\mathbf{r},\, t) = (\hbar/2\omega_q \rho_m V)^{1/2} \mathbf{e}_q [a_q \exp(i\mathbf{q} \cdot \mathbf{r} + q_z z) + HC], \tag{16}$$

where ω_q is the phonon frequency, ρ_m the mass density of the crystal, V the volume, e_q the polarization vector, a_q the phonon destruction operator, $\mathbf{q}$ the two-dimensional phonon wave vector ($x - y$ direction), q_z the component in the z direction, and HC the Hermitian conjugate. Using the golden rule and assuming equipartitioning [39], the scattering rate is given by

$$\frac{1}{\tau_{ac}} = \frac{\pi}{\hbar} \left(\frac{kT\Xi_1^2}{c_1^2 \rho_m V} \right) \sum_{q,q_z} B^2 \, \delta(E_k - E_{k\pm q} \mp \hbar\omega_q), \tag{17}$$

where τ_{ac} is the acoustic scattering time, the upper and lower signs stand for phonon absorption and emission, respectively, c_1 the velocity of sound, and

$$B = \int_0^\infty g(z) e^{iq_z z} \, dz. \tag{18}$$

As before, $g(z)$ is the square of the normalized z-dependent part of the wave function, which is $(2/L)\sin^2(\pi z/L)$ for a square well of length L. For inversion layers, the variational function of Stern and Howard [5], $g(z) = \frac{1}{2}b^3 z^2 \exp(-bz)$, where b is an adjustable parameter, is a good approximation. The value of B and especially the value of $\Sigma_{q_z} B^2$ depends on the form of the function $g(z)$ and, therefore, on the form of the potential well. If one converts the summation over q_z into an integration, then

$$\int_{-\infty}^{+\infty} dq_z \, B = \frac{1}{(d \cdot \text{const})}, \tag{19}$$

where d is some characteristic length of the potential. Inserting this into Eq. (17) and using $\rho d \cdot \text{const} = \rho_{2d}$, a two-dimensional mass density, one obtains

$$1/\tau_{ac} = kTm^*\Xi_1^2/\hbar^3\rho_{2d}c_1^2. \tag{20}$$

For the square well, $\rho_{2d} = \frac{4}{3}L\rho$, and for Stern and Howard's [5] approximation for $q(z)$, one has $\rho_{2d} \simeq 3.6d\rho$, with $d = 3/b$. More refined calculations including the anisotropy of the many-valley conduction-band structure have been performed [14–16]. However, present experimental results do not allow one to assess whether these calculations give improved estimates of the scattering rate. The calculation of all the other electron–phonon interactions via the deformation potential proceeds along similar lines and we give only the result for the matrix elements of the scattering probability or the scattering rates if they can be calculated explicitly.

Assuming that the elastic constants of SiO$_2$ and Si are sufficiently different and that the interface is sufficiently sharp, we would have scattering predominantly by Rayleigh waves. The lattice displacement is then [14]

$$\mathbf{u}_{\parallel}(\mathbf{r}) = \sum_q i\sqrt{N}\,\frac{q}{|q|}\left(e^{-\gamma qz} - \frac{2\gamma^n}{1+n^2}e^{-nqz}\right)(\mathbf{a}_q e^{i\mathbf{r}q} - \text{HC}),$$

$$\mathbf{u}_{\perp}(\mathbf{r}) = -\sum_q \mathbf{n}\sqrt{N}\left(\gamma e^{-\gamma qz} - \frac{2\gamma^n}{1-n^2}e^{-nqz}\right)(\mathbf{a}_q e^{i\mathbf{r}q} + \text{HC}), \qquad (21)$$

$$\mathbf{u}(\mathbf{r}) = \mathbf{u}_{\parallel}(\mathbf{r}) + \mathbf{u}_{\perp}(\mathbf{r}),$$

with

$$N = (\hbar/2S\rho c_R)(\gamma - n)(\gamma - n + 2\gamma n^2)/2\gamma n^2,$$

where c_R is the velocity of the Rayleigh wave, S the interface area, and γ and n functions of the Poisson ratio [14]. We then obtain for the matrix element (assuming $B \approx 1$)

$$M = -\Xi_{ac}\sqrt{N}q(1 - \gamma^2)(N_q + \tfrac{1}{2} \pm \tfrac{1}{2})\delta_{k\pm q},$$

For further details on acoustic mode scattering see, for example Refs. 14,39. It has not yet been assessed with certainty if and to what extent Rayleigh waves contribute [40], although n- and p-channels seem to behave differently. For the p-channel, Rayleigh waves may dominate; for the n-channel, the bulk phonons may scatter more strongly. If we consider scattering by bulk phonons only, then from the data of Fig. 3 we obtain a deformation potential constant of $\Xi_1 = 6$ eV, which is reasonable. Until now we considered only scattering within one subband, which is a good approximation for acoustic phonons since their energy is small.

2. Optical Phonons

In covalent semiconductors with two or more nonequivalent atoms per unit cell, optical modes corresponding to movements of these atoms relative to one another are allowed. These optical modes interact with the electrons to cause scattering. Although intravalley optical interaction is forbidden by symmetry in silicon, scattering between the different minima of the conduction band proceeds by essentially the same interaction [26,40], whether these minima are equivalent or not, provided the proper phonons for the intervalley umklapp process are used in the transition. The matrix element generally is found from a deformable ion model, and the lowest-order term that arises is the zero order, which is given by [40,41,42]

$$|\langle k \pm q|H_1|k\rangle|^2 = (D^2h/2V\rho_m\omega_0)|I|^2(N + \tfrac{1}{2} \pm \tfrac{1}{2}), \qquad (22)$$

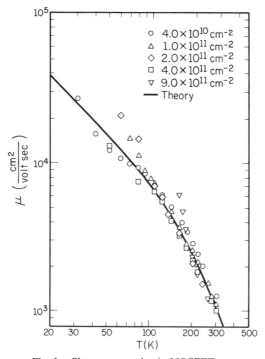

Fig. 3. Phonon scattering in MOSFETs.

where D is an effective energy shift per unit displacement (commonly given in terms of electron volts per centimeter), ω_0 the frequency of the relevant phonon mode, which is usually taken to be independent of the phonon wave vector for optical and intervalley processes, N the Bose–Einstein occupation factor for the phonons, ρ_m the mass density of the crystal, V the volume of the interaction space, $\hbar$ Planck's constant divided by 2π, the upper sign refers to the emission of a phonon by the electron, the lower sign refers to the absorption of a phonon by the electron, and the overlap integral is

$$|I| = \int_0^\infty \psi_{k'}^* \exp[(i\mathbf{q} + i\mathbf{g}) \cdot \mathbf{r}]\psi_k \, dV, \qquad (23)$$

where $\mathbf{g}$ is a vector of the reciprocal lattice. In quantized inversion levels, the motion parallel to the surface is nearly free-electron-like. For Stern's wave function [1,5], expressed as

$$\psi_k = \zeta_i(z) \exp(i\mathbf{k} \cdot \mathbf{r}), \qquad (24)$$

where $\mathbf{k}$ is the two-dimensional wave vector parallel to the surface, Eq.

(2) reduces to

$$|I| = \int_0^\infty \zeta_i(z)\zeta_j(z)\,dz, \qquad -k' + k + q + g = 0, \tag{25}$$

where the indices i, j refer to initial and final states of the electrons. In actuality, an index i or j refers to an index pair (μ, β), where μ designates the particular subband of the βth set of valleys. If one valley or one equivalent set of valleys are present, then β is an extraneous index.

Using the normal definition of the scattering rate, the probability per unit of a carrier being scattered out of k is given by [25]

$$\frac{1}{\tau} = \frac{2\pi}{\hbar} \sum_q \frac{D^2\hbar}{2V\rho_m\omega_0} I_{ij} \left(N + \frac{1}{2} \pm \frac{1}{2}\right) [1 - f_0(E')]\,\delta(E' - E \pm \hbar\omega_0), \tag{26}$$

where $f_0(E')$ is the distribution function, the term in the square brackets allows for the possibility of the final state being occupied, and the δ function ensures the conservation of energy. In Eq. (26), no separation of the two terms for emission and absorption is included, although this will be done later. Since the matrix element is independent of q, the sum over phonon states can be taken into a sum over final energy states as [43]

$$\sum_q \delta(E' - E \pm \hbar\omega_0) \to \int \delta(E' - E \pm \hbar\omega_0)\,\rho(E')\,dE', \tag{27}$$

where $\rho(E')$ is the density of final states per unit energy. The density of states for a particular subband and valley is just [1]

$$\rho(E) = S(m_{1\beta}m_{2\beta})^{1/2}/\pi\hbar^2, \tag{28}$$

where S is the surface area of the two-dimensional set of states. Equations (26) and (28) can readily be combined and a sum carried over all subbands and valleys to give a net scattering rate of

$$\frac{1}{\tau} = \frac{D^2}{\rho_m\hbar^2\omega_0} \sum_{\alpha,\beta} I_{ij} \frac{(m_{1\beta}m_{2\beta})^{1/2}}{w_\alpha} \{Nu_0(E - E_\alpha + \hbar\omega_0)[1 - f_0(E + \hbar\omega_0)]$$

$$+ (N + 1)u_0(E - E_\alpha - \hbar\omega_0)[1 - f_0(E - \hbar\omega_0)]\}. \tag{29}$$

In Eq. (29) the rates for emission and absorption have been separated and it is to be remembered that E_α and the level width w_α in the z direction refer to the αth subband of the βth set of valleys. Multiplicity of the final state valleys must also be taken into account in Eq. (29). The function $u_0(x)$ is the unit step function, defined by

$$u_0(x) = \begin{cases} 1, & x \geq 0, \\ 0, & x < 0. \end{cases} \tag{30}$$

The only energy dependence in Eq. (29) is tied to the distribution function $f_0(E)$ and the unit step functions $u_0(E - E_\alpha \pm \hbar\omega_0)$. The explicit energy dependence can be removed by taking an average over the distribution function.

In many applications, a situation arises in which the zero-order process just described is forbidden by the symmetry of the states involved [44–46]. This has been discussed for the case of bulk silicon [47], and relaxation times have been determined for bulk material calculated for scattering via the first-order interaction. In this section, the relaxation time for scattering via the first-order interaction will be calculated for the quantized inversion layer.

The matrix element for the first-order interaction was found by Harrison [40], and is given by

$$|\langle k \pm q|H_1|k\rangle|^2 = (D_1{}^2\hbar q^2/2V\rho_m\omega_0)(N + \tfrac{1}{2} \pm \tfrac{1}{2})I_{ij}, \tag{31}$$

where D_1 is the effective energy shift, or deformation potential, for the optical interaction. The probability per unit time of a carrier being scattered out of k is given by

$$\frac{1}{\tau'} = \frac{2\pi}{\hbar} \sum_q \frac{D_1{}^2\hbar q^2}{2V\rho_m\omega_0} I_{ij}(N + \tfrac{1}{2} \pm \tfrac{1}{2}) \, \delta(E' - E \pm h\omega_0)[1 - f_0(E')], \tag{32}$$

where τ' is used to signify the first-order interaction. Although the matrix element is dependent on q, the sum over final states can still be carried out as in Eq. (27), since

$$q^2 = |k \pm k'|^2 = k^2 + k'^2 \pm 2kk' \cos\theta \tag{33}$$

in two dimensions, which becomes, for this case,

$$\sum_q q^2 \, \delta(E' - E \pm \hbar\omega_0)$$
$$\rightarrow \iint (k^2 + k'^2 \pm 2kk' \cos\theta) \, \delta(E' - E \pm \hbar\omega_0) \, \rho(k') \, dk'. \tag{34}$$

The integral over dk' involves an integration over the planar angle θ, for which the term in $\cos\theta$ integrates to zero. The remainder of the integration is straightforward and can be carried out in E' space. The sum over q then yields just the factor

$$\rho(E \pm \hbar\omega_0)(2k^2 + m\omega_0/\hbar), \tag{35}$$

where $m = (m_1m_2)^{1/2}$ is the appropriate density of states mass for the subband involved. This can be rewritten as

$$\rho(E \pm \hbar\omega_0)[(2m/\hbar^2)(2E \pm \hbar\omega_0)]. \tag{36}$$

The term in the density of states is just the same result as from Eq. (27). The added factor in the square brackets stems from the nature of the first-order interaction and is the same added factor that occurs for the three-dimensional, first-order treatment [47]. Equations (32) and (36) can then be combined to give a net scattering rate of

$$
\frac{1}{\tau'} = \frac{2D_1^2}{\rho_m \hbar^4 \omega_0} \sum_{\alpha,\beta} I_{ij} \frac{m_{1\beta} m_{2\beta}}{w_\alpha} \{N(2E + \hbar\omega_0)u_0(E - E_\alpha + \hbar\omega_0)
$$
$$
\times [1 - f_0(E + \hbar\omega_0)] + (N + 1)(2E - \hbar\omega)
$$
$$
\times u_0(E - E_\alpha - \hbar\omega_0)[1 - f_0(E - \hbar\omega_0)]\}, \qquad (37)
$$

where again it is to be remembered that E_α and w_α refer to the αth subband of the βth set of valleys. The explicit energy dependence can be removed by taking an average over the distribution function.

3. Remote Optical Phonons

In the preceding discussion we mentioned the main scattering mechanisms one would expect in Si. We have to consider, however, that the charge carriers are at a distance of 10^{-6} cm or less from the strongly polar SiO_2. Therefore, the "fringing fields" of the polar interface modes can also scatter the charge carriers [48]. For a free surface, the macroscopic electric fields outside the crystal and the scattering rate of electrons due to the polar surface modes were calculated by Wang and Mahan [49]. Their treatment was generalized to the case of two adjacent dielectrics and size quantization of inversion-layer carriers [38], and the scattering rate due to this remote polar interface phonon was found to be

$$
\frac{1}{\tau \rho_0} = \frac{e^2 \omega_1}{4\pi\epsilon_0} \left[\frac{1}{\epsilon_\infty + \epsilon_{Si}} - \frac{1}{\epsilon_{st} + \epsilon_{Si}} \right]
$$
$$
\times \int_0^\infty \int_0^{2\pi} \left(N_q + \frac{1}{2} \pm \frac{1}{2} \right) \delta(E_{k\pm q} - E_k \pm \hbar\omega_1) \, dq \, d\phi, \qquad (38)
$$

where ϵ_0, ϵ_∞, ϵ_{st} and ϵ_{Si} are the dielectric constants of free space, the optical and the static dielectric constant of SiO_2, and the dielectric constant of silicon, respectively. To arrive at this formula it was assumed that $|q|d \ll 1$. For larger q vectors, the interaction decreases rapidly because the interface mode decays exponentially with $|q|d$ [38], where, as outlined before, d is the average distance of the electrons from a hypothetical interface plane. Since in some cases $d \approx 10$ Å, it is certainly necessary to include a realistic range for this parameter.

For the case of two polar optical neighboring media, the term $\omega_1[(1/\epsilon_\infty +$

$\epsilon_{Si}) - (1/\epsilon_{st} + \epsilon_{Si})]$ in Eq. (38) has to be replaced by

$$\frac{1}{\omega_{\pm}}\left[\frac{\omega_{TO1}^2(\epsilon_{st1} - \epsilon_1)}{(\omega_{\pm}^2 - \omega_{TO1}^2)^2} + \frac{\omega_{TO2}^2(\epsilon_{st2} - \epsilon_{\infty2})}{(\omega_{\pm}^2 - \omega_{TO2}^2)}\right]^{-1},$$

where $\omega_{\pm}$ are solutions of

$$(\omega^2 - \omega_{TO1}^2)(\omega^2 - \omega_{LO2}^2) = -\frac{\epsilon_{\infty1}}{\epsilon_{\infty2}}(\omega^2 - \omega_{TO2}^2)(\omega^2 - \omega_{LO2}^2). \qquad (39)$$

Here $\omega_{TO,i}$ ($i = 1,2$) is the angular frequency of the transverse polar optical mode, ω_{LOi} the longitudinal frequency, and ϵ_{sti}, $\epsilon_{\infty i}$ the static and optic dielectric constants.

These are the scattering mechanisms as they are presently known. We will apply them to calculate the current–voltage characteristics of "ideal" MOS transistors with very short channels in a following section. From this detailed study, we will then deduce simple formulas that reflect the basic physics of small MOS devices.

III. THE SILICON SURFACE SYSTEM

A. Level Structure

In a quantized electron inversion layer at the surface of (100)-oriented silicon, the six equivalent minima of the bulk silicon conduction band split into two sets of subbands. One set consists of the subbands arising from the two valleys that show the longitudinal mass in the direction perpendicular to the surface. This set has energy levels E_0, E_1, E_2, ..., in the notation of Stern and Howard [5]. The subband E_0 that is lowest at the surface belongs to this set. The other set of subbands arises from the four equivalent valleys that show a transverse mass in the direction normal to the surface. This set has energy levels designated as E_0', E_1', E_2', ... Generally, these levels line up such that E_0' is almost degenerate with E_1. The projections of the valley onto the [001] plane and the various subbands are shown in Fig. 4. Except at very low temperatures, it is unreasonable to assume that all of the electrons are in the lowest subband. However, it is not a bad approximation to assume that most of the electrons occupy the three lowest levels E_0, E_1, and E_0', and this is usually done. We shall also take this approximation in the following calculations of transport within the inversion layer. Scattering within and between the various subbands involves the entire range of intervalley phonons as well as acoustic phonons and surface scattering.

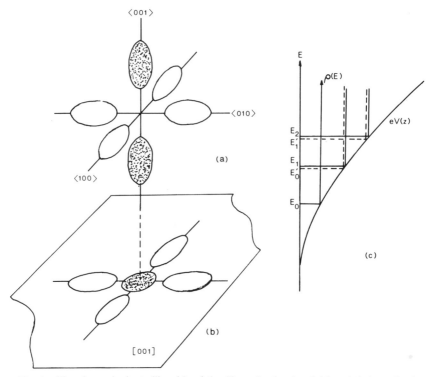

Fig. 4. The six equivalent ellipsoids of the Si conduction band (a) and their projection onto the [001] plane (b), which leads to two sets of subbands (c).

The values for the various masses are their normal values and are tabulated in Stern and Howard [5]. The level thicknesses w_j are taken from the self-consistent calculations reported in the same references and from estimates made using a triangular-well approximation. Their variations as well as the variations of the energy levels with temperature of the lattice are explicitly taken into account.

In Fig. 5 the average width of the inversion layer and the fraction of carriers in the lowest subband are shown for various temperatures for $n_s > N_{depl} = (N_d - N_a)d$, where d is the depletion width. From Fig. 5, it is apparent that as n_s is increased, the surface field F_s increases, causing a deeper potential well. Thus the subbands are moved further apart, causing the higher lying subbands to be emptied of carriers.

A word of caution should be entered at this point. Figure 5 is calculated using just the simple Hartree approximation leading to the equations of

Section I.B. Many-body effects, especially at low temperatures, can modify the actual energy levels, which would affect the results of Fig. 5. However, at 300 K, these effects are not expected to be large, but this assumption has not been verified as yet.

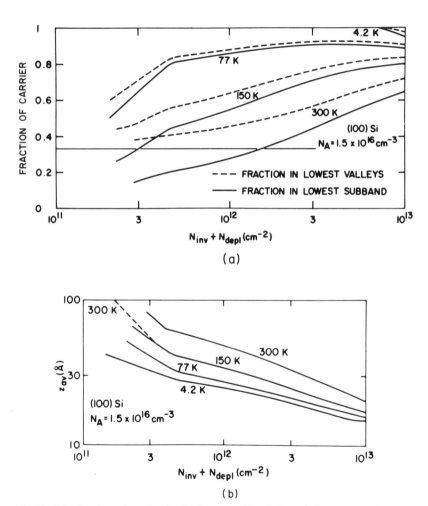

Fig. 5. The fraction of carriers in the lowest subband (a) and the average width of the quantized layer (b). See the text for a discussion of these levels. Reprinted with permission from F. Stern, *Crit. Rev. Sol. State Sci.* **4**, 499 (1974). Copyright The Chemical Rubber Co., CRC Press, Inc.

B. Contributing Phonons

The conduction band of silicon is characterized by six equivalent minima located along Δ approximately 83% of the way to X from the Γ point. Intervalley scattering can proceed by two different groups of phonons. One set, denoted g phonons, couples one valley to its partner along the $\langle 100 \rangle$ axis. The other set, denoted f phonons, couples this valley to the four valleys lying on the $\langle 010 \rangle$ and $\langle 001 \rangle$ axes. Because the valleys lie more than halfway to X, all of these processes involve umklapp processes. The net g-phonon wave vector corresponds to $0.34X$, while the f phonons arise from phonons whose wave vector lies along Σ extended along S in the plane perpendicular to the [100] axis. These latter phonons possess Σ symmetry, but they lie very near the X point so that their energies are very near to those of the X point [44,50]. Because of the multitude of phonons possible, Long [50] used just two equivalent phonons to treat transport in silicon over the temperature range 30–350 K. One of these was high-energy, 630-K phonon used to represent the effect of the LO + TO intervalley processes and the LO intravalley process. The other was a 190-K phonon used to represent the effect of the LA + TA interactions. Not considering the possibility that many of these processes were forbidden, he determined values for the coupling constants by fitting the experimental data for various transport measurements. In doing so, he found that the low-energy process was considerably weaker than the high-energy process, with the two contributing in the ratio 0.15/2. Subsequently, Streitwolf [45] and Lax and Birman [46], in considering the crystal symmetry of silicon, pointed out that only the $\Delta_2'(\text{LO})$ phonon was allowed for the g-phonon interaction, and the $\Sigma_1(\text{LA, TO})$ was allowed for the f-phonon interaction. Both of these processes would contribute to Long's 630-K phonon. All of the processes that could contribute to the low-energy, 190-K phonon are, in fact, forbidden by symmetry. Norton *et al.* [51], after noting the constraints introduced by symmetry, recalculated the transport properties, but they still were required to include, ad hoc, the low-energy phonon in order to fit the experimental data. The most explicit evidence for the presence of the phonons is found in the magneto-phonon resonance data of Portal *et al.* [52]. Magnetophonon resonance occurs when the Landau-level separation is a submultiple of the phonon energy, or $\hbar\omega_0 = n\hbar\omega_c$, where n is an integer and $\omega_c = eB/m$, the cyclotron resonance frequency. In the measurements of magnetophonon resonance in silicon, oscillatory series were observed that corresponded to most of the phonons mentioned at the start of this section, and the relative strengths of the various phonons were estimated to be near that assumed by Long. Although the low-energy phonons are explicitly forbidden in the zero-order interaction, their observation in experimental measurements suggests that they are interacting via a first-order interaction [47]. The occurrence of this interaction via a first-order process would also explain why the low-energy process was found to be so weakly coupled.

Scattering between the two equivalent valleys in the E_0 and E_1 subbands involves g-type phonons. These phonons couple the $\langle 001 \rangle$ valley with its partner along the $\langle 00\bar{1} \rangle$ direction. In the present calculation, the scattering between these two minima will be treated by using a high-energy phonon of 750-K equivalent activation temperature ($T_a = \hbar\omega_0/k_B$) and a low-energy phonon of 134-K equivalent activation temperature. The former is chosen as it is the value of the LO phonon that is allowed in the g process. It is treated via a zero-order interaction. The 134-K phonon is found to be the strongest low-energy g phonon in magnetophonon resonance studies in silicon [52]. This phonon is forbidden in zero order and must be treated by a first-order interaction.

Scattering between the two subbands, E_0, E_1, and the four E_0' subband valleys involves f-type phonons. For this scattering, which is primarily responsible for repopulation among the subbands, f phonons of 630- and 230-K equivalent activation temperatures are utilized. The former is zero-order coupled while the latter is the value of the zone-edge TA phonon, which is forbidden to zero order. This is treated via the first-order interaction.

Scattering between the valleys of the E_0' subband involves both the g- and f-type phonons. It is a viable procedure to use just two effective phonons of 630 and 190 K for these phonons, as is also done in treating bulk silicon. The former is zero-order coupled while the latter is first-order coupled since all phonons that could contribute to this low-temperature interaction are forbidden. In this work, all of the high-energy phonons are assumed to be coupled with a value of $D = 9 \times 10^8$ eV/cm and all of the first-order coupled phonons are assumed to be coupled with $D_1 = 5.6$ eV. A value of $\Xi_1 = 6$ eV, appropriate to the surface, is taken for the deformation potential of acoustic scattering.

C. The Balance Equations

As previously mentioned, although it is unreasonable to assume that all of the electrons are in the lowest subband, it is not a bad approximation to assume that most of the electrons occupy the three lowest levels E_0, E_1, and E_0'. This approximation will generally be taken here in regard to the hot electron problem. As just discussed, the valleys are uncoupled in the balance equations and these can be solved for each valley separately to determine its parameters. Then the fractional occupations of the different valleys are found from the transition rate equations used in the population balance equation and the total effective mobility or velocity is found. The various balance equations within a single level are given as

$$eF = \langle dp/dt \rangle = m_c v_d \langle 1/\tau \rangle, \tag{40}$$

$$ev_d F = -\langle dE/dt \rangle, \tag{41}$$

for momentum p and energy E, with m_c the conducting mass and v_d the drift velocity and

$$\frac{dn_i}{dt} = -\sum_j \frac{\partial n_i}{\partial t}\bigg|_{i \rightarrow j} + \sum_j \frac{\partial n_i}{\partial t}\bigg|_{j \rightarrow i} \tag{42}$$

for the population, where the terms on the right represent collisional transitions between levels and the sums run over all levels $j \neq i$.

One can question the use of only three subbands, since at these temperatures many more could be expected to be occupied. However, the role of other subbands is not expected to greatly affect the results. Although the higher subbands show higher mobilities due to an increase of the effective level thickness, the carriers in these levels get hot at lower electric fields, so that the net effect of these levels does not modify to any great extent the high-field behavior, which is found from treating only three subbands.

The balance equations given by Eqs. (40)–(42) are obtained by taking moments of the Boltzmann transport equation, assuming that the distribution function in each valley is a drifted Maxwellian:

$$f(\mathbf{k}) = C \exp[-\hbar^2(\mathbf{k} - \mathbf{k}_d)^2/2m_c k_B T_e], \tag{43}$$

where $\hbar k_d = m_c v_d$ is the average drift momentum of the electron gas, C a constant for normalization purposes, and T_e the electron temperature. From Eq. (43), it follows that

$$f_0(E) = C \exp(-E/k_B T_e) \tag{44}$$

and

$$f_1(E) = v(m_c v_d/2k_B T_e)f_0(E). \tag{45}$$

The actual moments themselves are complicated by the fact that the inversion layer is describable by multilevels and hence is a coupled system. We must account for the repopulation effects that can also occur.

The Boltzmann transport equation is given as

$$\frac{\partial f}{\partial t} + \mathbf{v} \cdot \nabla f + e\mathbf{F} \cdot \nabla_p f = \sum_{p'} \{W(\mathbf{p}, \mathbf{p}')f(\mathbf{p}') - W(\mathbf{p}', \mathbf{p})f(\mathbf{p})\}, \tag{46}$$

where $W(\mathbf{p}, \mathbf{p}')$ is the scattering rate from state $\mathbf{p}'$ to state $\mathbf{p}$. In the following, we shall ignore the inhomogeneity term $\mathbf{v} \cdot \nabla f$. We define the average of $\phi(\mathbf{p})$ as

$$\langle \phi(\mathbf{p}) \rangle = \frac{1}{n} \int \phi(\mathbf{p}) f(\mathbf{p}) \, d\mathbf{p}. \tag{47}$$

To begin, we multiply Eq. (45) by an arbitrary function of $\phi(\mathbf{p}_i)$, assuming $\phi(\mathbf{p}_i)$ and $f_i(\mathbf{p}_i)$ represent the ith valley (or subband in this case). Then integration over $\mathbf{p}_i$ yields

$$\frac{\partial}{\partial t} n_i \langle \phi \rangle_i - n_i e \mathbf{F} \cdot \langle \nabla_{\mathbf{p}_i} \phi \rangle_i = \sum_{\mathbf{p}_i} \{ \sum_{\mathbf{p}_i'} [W(\mathbf{p}_i, \mathbf{p}_i') f(\mathbf{p}_i')$$

$$- W(\mathbf{p}_i', \mathbf{p}_i) f(\mathbf{p}_i)] + \sum_{\mathbf{p}_j'} [W(\mathbf{p}_i, \mathbf{p}_j') f(\mathbf{p}_j')$$

$$- W(\mathbf{p}_j', \mathbf{p}_i) f(\mathbf{p}_i)]\} \phi(\mathbf{p}_i). \tag{48}$$

Here, we have separated the intravalley and intervalley contributions to the scattering processes. When the initial and final states lie in the same valley, they are describable by the same local distribution function. With this in mind, the intravalley terms on the right-hand side of Eq. (47) (summation over p_i') can be treated by a simple change of variables and this term becomes

$$n_i \langle \Gamma_\phi(\mathbf{p}_i) \rangle_i, \tag{49}$$

where

$$\Gamma_\phi(\mathbf{p}_i) = \sum_{\mathbf{p}_i'} W(\mathbf{p}_i', \mathbf{p}_i)[\phi(\mathbf{p}_i') - \phi(\mathbf{p}_i)]. \tag{50}$$

If $\phi = C$, this term vanishes and, as expected, makes no contribution to a density balance equation such as Eq. (42).

The intervalley (or intersubband) terms can be written as

$$n \left\langle \frac{d\theta}{dt} \bigg|_{\text{coll}} \right\rangle = \sum_{\mathbf{p}_i} \phi(\mathbf{p}_i) \sum_{\mathbf{p}_j'} \{ W(\mathbf{p}_i, \mathbf{p}_j') f(\mathbf{p}_j')$$

$$- W(\mathbf{p}_j', \mathbf{p}_i) f(\mathbf{p}_i) \}$$

$$= \sum_{\mathbf{p}_j'} f(\mathbf{p}_j') \phi(\xi_i) \sum_{\mathbf{p}_i} W(\mathbf{p}_i, \mathbf{p}_j')$$

$$- \sum_{\mathbf{p}_i} f(\mathbf{p}_i) \phi(\mathbf{p}_i) \sum_{\mathbf{p}_j'} W(p_j', \mathbf{p}_i), \tag{51}$$

where we have used the energy conserving δ function inherent in

$W(\mathbf{p}_i, \mathbf{p}_j')$ to define ξ_j through the relation

$$\epsilon_j' = \epsilon_i \pm \hbar\omega_0, \tag{52}$$

so that ξ_j is a function of $\mathbf{p}_j'$. Introducing the scattering rate as

$$\Gamma(\mathbf{p}) = \sum_{\mathbf{p}'} W(\mathbf{p}', \mathbf{p}), \tag{53}$$

Eq. (50) becomes

$$n\langle d\phi/dt|_{\text{coll}}\rangle = n_j\langle \phi(\xi_j')\Gamma(\mathbf{p}_j')\rangle_j - n_i\langle \phi(\mathbf{p}_i)\Gamma(\mathbf{p}_i)\rangle_i. \tag{54}$$

By inserting the details of the various scattering processes, the individual moment equations can be readily set up. However, because of the multiplicity of scatterers, the individual equations are quite complicated. We shall not delve deeper into their structure here. The details have been treated elsewhere [4,53].

D. Mobility

The calculated low-field ohmic mobilities for electrons are 600 cm²/V sec at 300 K and 2500 cm²/V sec at 77 K for an inversion density of $n_s = 10^{12}/\text{cm}^2$. For $n_s = 10^{13}/\text{cm}^2$, a mobility at 300 K of 390 cm²/V sec is cal-

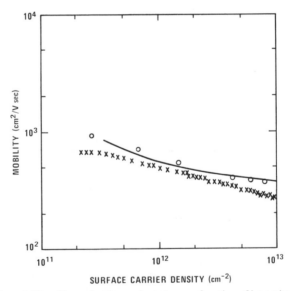

Fig. 6. The mobility of inversion-layer electrons as a function of inversion charge density at 300 K on (100) Si. The data are representative of that of Fang and Fowler [3,11], Sato *et al.* [54], and Sabnis and Clemens [55].

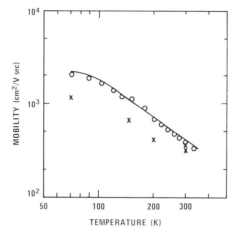

Fig. 7. The mobility of inversion-layer electrons as a function of temperature for $n_s =$ $6 \times 10^{12}/cm^2$. The data are the same as those of Fig. 6.

culated. These values are sensitive to the value of surface roughness and surface charge. Here we use the values of Δ and L as previously discussed. This gives the mobility shown in Figs. 6 and 7. In Fig. 6, we show the mobility as a function of inversion-layer density, while in Fig. 7 we show the mobility as a function of temperature. The dominant scattering process is surface-roughness scattering, as discussed in the previous section. Representative data of Fang and Fowler [3,11], Sato et al. [54], and Sabnis and Clemens [55] are shown for reference.

E. High-Electric-Field Effects

In Fig. 8, the velocity is shown as a function of electric field at 300 K for $n_s = 10^{12}/cm^2$ and $n_s = 10^{13}/cm^2$ for (100) silicon. The mobility increases above the ohmic value for fields in the range $10^2 - 10^3$ V/cm, due primarily to the fall off of local potential scattering as T_e increases. For fields above 10^3 V/cm, the lower subbands become very hot and carrier transfer into the E_0' valley begins. This upper valley remains relatively cool for fields below 20 kV/cm. For fields above this, however, heating of this upper set of valleys begins to become significant and the repopulation will eventually redistribute the carriers equally among all valleys at very high fields. The carrier transfer from the E_0 and E_1 valleys to the E_0' valley can be seen in the velocity–field curves as an inflection point. It is most prevalent in Fig. 9 in the curve for $n_s = 10^{13}/cm^2$ at a field in the vicinity of 10^4 V/cm. The relative populations of the E_0 and E_0' valleys are shown in Fig. 9.

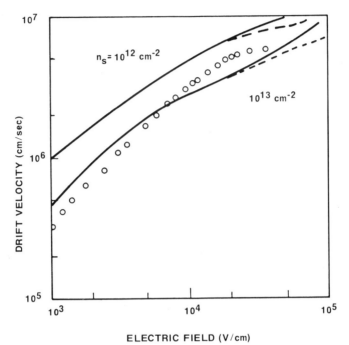

ELECTRIC FIELD (V/cm)

Fig. 8. Velocity–field curves calculated for $n_s = 10^{12}/\text{cm}^2$ and $n_s = 10^{13}/\text{cm}^2$, on (100) silicon at 300 K are shown. The data of Fang and Fowler [3,11] for $n_s = 6.6 \times 10^{12}/\text{cm}^2$ are shown for comparison. The solid curves are for the drain–source electric field parallel to the [011] direction, so that the four valleys of the E_0' subband make the same angle with the field and are thus equivalent. The dashed curves are for the drain–source field parallel to the [010] direction, for which repopulation occurs among the four valleys of the E_0' subband as it becomes hot.

The carrier transfer from the E_0 to the E_0' valley can be seen in the velocity–field curves as an inflection point. It is most prevalent in Fig. 9 for fields in the range 10–20 kV/cm. The experimental data of Fang and Fowler [3,11] for $n_s = 6 \times 10^{12}/\text{cm}^2$ is also shown, and the agreement is relatively good. As can also be seen, when the upper valley begins to get hot, the direction of the field is important, since all four valleys are equivalent only for a field along the (110) direction. For other field directions, redistribution among these valleys becomes important.

The repopulation among the various subbands has an analog in bulk silicon in which repopulation among the (100) valleys occurs for an electric field parallel to one of the (100) directions [56]. In this case, not all of the valleys are equally heated by the field. If the preceding calculations are repeated at lower temperatures, it is found that the field at which E_0 to E_0' transfer occurs decreases with the temperature. At 77 K, repopulation occurs over a relatively narrow range of electric field and is essentially complete by a field of 300 V/cm. Even though the large population shift occurs from the light-mass E_0 valley to the heavy-mass E_0' valley, the velocity increases monotonically with electric field. No negative differential conductivity exists in this system, at least for the values utilized in the calculations. It would indeed be surprising if such an effect did occur since the ratio of the densities of states for these two sets of valleys is so low, being essentially only about 4. The value of the field for the onset of repopulation is only 200 V/cm and this agrees well with that found in the bulk.

Experiments have also shown negative differential conductivity at 4.2 K in high-purity samples [57] and at higher temperatures in devices with high levels of oxide charge [58]. In these experiments, the NDC was observed at very high electric fields, well above those expected for repopulation among the various subbands. At these fields, the carriers are very hot, and the inversion layer is expected to be nearly bulklike [4,23]. In bulk material, no NDC has ever been seen other than for equivalent valley repopulation, so that these observations remain an unexplained enigma. Only weak effects are expected from remote polar phonons, but this weak effect may be present in recent experimental results [59].

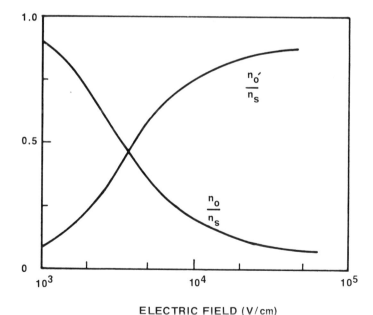

Fig. 9. Relative populations of the E_0 and E_0' subbands are shown as a function of electric field for $n_s = 10^{13}/cm^2$ at 300 K on (100) Si.

F. High-Frequency Effects

Let us now turn our attention to the ac or small-signal, microwave conductivity of the inversion layer. As previously mentioned, for high electric fields and hot-electron conditions, the transport becomes very nearly bulk-like. One might then expect to obtain a microwave conductivity that is just representative of the hot-electron microwave conductivity in bulk material [60]. There is one situation in which this is not the case for the microwave conductivity. As we have seen, for modest electric fields, repopulation of the various subbands occurs, with carrier transfer occurring from the lower, light-mass subbands to the higher-lying, heavy-mass subbands. For applied dc fields in this electric field range, differential repopulation is induced by the microwave fields and this effect contributes substantially to the observed microwave conductivity [61]. In Fig. 10, the microwave conductivity is shown as a function of frequency for two values of the source–drain electric field. The parameters are the same as those for which the velocity–field curve of Fig. 8 was calculated. In Fig. 10, the effect of differential repopulation can be seen as the difference between the dashed and solid curves, the latter including differential repopulation in the presence of the ac field. It can be seen that the differential repopulation arising from the subbanding makes a sizable contribution to the microwave conductivity and, in fact, lowers the microwave conductivity. The peak in the conductivity near 10^{12} Hz is significant in that it indicates that for electric fields in the range necessary for repopulation, transient drift velocities would exhibit overshoot effects.† That is, if an electric field of 20 kV/cm, for example, were switched on at $t = 0$, the velocity would rise to a value characteristic of the light-mass subbands. Then, as repopulation occurred, the velocity would settle to its lower steady-state value. From the figure, it is evident that this transient overshoot would last for 3 to 5 psec. In short, such effects arise primarily because the energy and momentum balance are achieved on somewhat different time scales.

G. Miscellaneous High-Field Effects

A topic of considerable importance in integrated circuit development and reliability is electron injection and trapping in silicon dioxide. Changes in the IGFET characterization caused by hot-electron injection into the oxide (or in tunneling states close to the interface) were docu-

---

† The overshoot velocity is well known in the III–V compound semiconductors (GaAs, InAs, etc.) and is discussed in Chapter 6, Volume 1 for the case of Si.

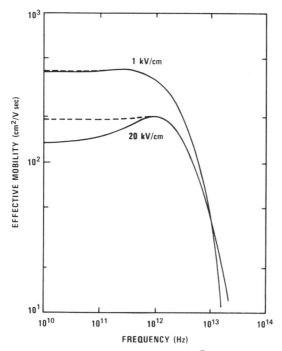

Fig. 10. Total effective ac mobility of the carriers in the inversion layer as a function of frequency for two different electric fields for $n_s = 6 \times 10^{12}/cm^2$ at 300 K on (100) Si. The dashed curves show the differences that result if the differential repopulation is neglected.

mented by Abbas and Dockerty [62], Hess *et al.* [58], and in numerous detailed and comprehensive accounts by, e.g., Ning [63]. Recently Forbes *et al.* [64] showed that these effects are accelerated and aggravated at low temperatures, possibly precluding any advantage to be gained from operation of IGFET circuits at low temperatures. The basic principle underlying all these effects is that electrons can escape from potential wells when they are accelerated by high electric fields and become "hot." A mechanical analogy is provided by the example of a ball rolling down a chute. The ball will stay in the chute if its kinetic energy remains small. However, if the ball gains adequate kinetic energy, then an obstacle can scatter the ball out of the chute. We can illustrate the basic physics by means of a buried-channel charge-coupled device [65]. The structure of an *n*-channel BCCCD is shown in Fig. 11a. An *n*-type top layer is obtained by ion-implanting arsenic with concentration N_1 into a *p*-type substrate. The potential profile of this device is shown in Fig. 11b. The dashed curve shows the potential profile with the application of a high gate voltage V_G in the absence of signal charge. A potential energy minimum for electrons is

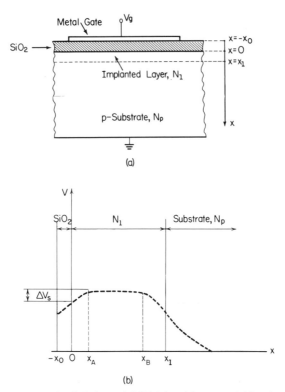

Fig. 11. A buried-channel CCD (a) and its potential barrier (b).

created within the region $x_A \leq x \leq x_B$, where electrons can be stored. The depth of the well containing the stored electrons is reduced as shown by the dashed curve. It is clear that electrons in the potential well can gain kinetic energy from high electric fields parallel to the interface. If they get extremely hot, a transition of almost all carriers from the potential well to the interface can occur, i.e., we have a field-induced transition of buried channel to surface channel. This is, of course, unwelcome and the device has to be designed in a way to avoid this effect. This imposes limitations on the charge-handling capacity of a BC-CCD.

The maximum charge-handling capacity implies the condition that the surface potential barrier $\Delta V = V_{ps} - V_{ss}$ is large enough to repel the electrons and to avoid the interface trapping. The spatial distribution of electrons in the potential well perpendicular to the interface is determined by the formula

$$n = n_0 \exp(e\phi/kT_c),$$

where ϕ is the potential and T_c the carrier temperature, which cannot generally be defined in the presence of high electric fields, since the energy distribution function can be highly non-Maxwellian for lower carrier concentrations.

In Fig. 12, the average carrier energy is plotted for the two limiting cases assuming a Maxwellian distribution function (strong carrier–carrier scattering) and using a distribution function obtained from Monte Carlo calculations. Using these values, one estimates a reduction of the charge-handling capacity by a factor of 2 for BC-CCDs with 10^{-4}-cm gate length and 5-V clock voltage swing. The conditions on device design are even more restrictive if one needs to avoid the inclusion of interface noise and interface capture (transfer inefficiency).

An entire set of high-field effects, which is not included in contemporary hot-electron treatments, is connected with inhomogenities in the electric field. Next a discussion of some of the more important effects will be given. In the current description of IMPATT devices it is always assumed that the avalanche zone is confined to the region of extremely high electric field. In the language of the electron-temperature model, this means that in this narrow region the temperature of the electrons is 10^4 K, but only 0.1 μm away the field is low and the electron temperature is 300 K. This huge temperature gradient is, of course, inconceivable. The electrons will carry energy with them and broaden by diffusion and drift the

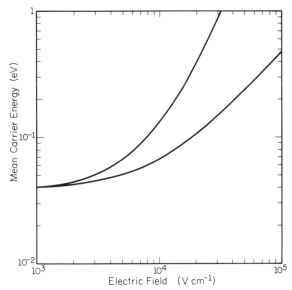

Fig. 12. Average carrier energy as a function of the field for a Maxwellian (upper curve) and by Monte Carlo (lower curve) techniques.

temperature profile, i.e., we have to include in physical models the electronic heat conduction. As a consequence the power-balance equation [Eq. (41)] has to be replaced by

$$\frac{\partial(nE)}{\partial t} = \mathbf{j} \cdot \mathbf{F} + n \left\langle \frac{dE}{dt} \right\rangle + \frac{\partial}{\partial x} \left\{ \kappa \frac{\partial T_e}{\partial x} + \frac{j}{e} \frac{\langle \tau E^2 \rangle}{\langle \tau E \rangle} \right\}. \tag{55}$$

This equation is valid for a spherical, single-valley band and was written in a one-dimensional form for simplicity. The thermal conductivity κ of the electron is given by $\kappa = 2kDn$. For deformation potential scattering ($\tau \simeq 1/E$), one obtains

$$\langle \tau E^2 \rangle / \langle \tau E \rangle = 2kT_c.$$

The effect of the thermal conduction of the free carriers is to broaden the T_c curves. Experimental evidence indicates that, indeed, in the range of high power, the avalanche zone spreads beyond the region of extremely high field.

The effect has also been shown to be important in charge-coupled devices [66]. Figure 13 shows typical fields under the gate of a surface-channel CCD and the corresponding electron temperature calculated with and without the inclusion of electronic heat conduction. One can see that the effect is pronounced. This will influence the transfer speed and efficiency of surface- and buried-channel, charge-coupled devices.

Hot-electron diffusion is important to a lesser extent in devices than is hot-electron drift. Therefore, we would like to point out only a few features of this effect. The power input to the electron gas is equal to the product vF [Eq. (41)]. Since diffusion and drift current can compensate each other, there exists the possibility of having very high electric fields present without average heating of the charge carriers, as the net current is zero. This is the case in a p–n junction without bias when the built-in field is high. Therefore, as soon as diffusion becomes important, the mobility and conductivity are no longer unique functions of the electric field but are only unique functions of the carrier temperature T_c, which in turn is a function of vF, not of F alone.

This makes calculations very complicated. A simplified approach was given by Hess and Sah [66]. Hot-electron diffusion is usually not included in device modeling because of the large numerical difficulties and in some cases because of the wrong assumption that the diffusion constant decreases with the electric field just as the mobility does. Also, the diffusion constant becomes dependent on the space coordinates and therefore a treatment of hot-electron diffusion as proposed by Stratton [67] (or an extension to explicitly space-dependent scattering rates) is necessary.

Two further problem areas must be mentioned. Because of the high cur-

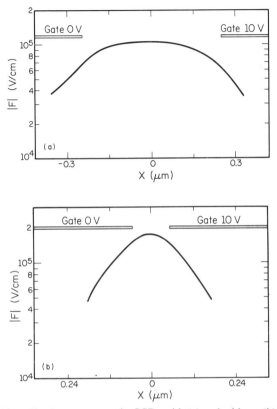

Fig. 13. Field profiles between gates in CCDs with (a) and without (b) the inclusion of electron heat conduction.

rent densities in some fine-line devices, a large density of lattice vibrations is generated in a very small volume. As a consequence, phonons can contribute appreciably to device performance. This has been shown for quantum-well, heterojunction lasers [68] but certainly can be important for transport properties in submicron devices as well.

Second, as the electrons (holes) acquire very high kinetic energies, they can lose their energy to an electron in a trap on the valence band, which then makes a transition to the conduction band. (This is exactly the inverse Auger process, where an electron makes a transition from the conduction band to the valence band and gives the energy to another electron in the conduction band.) In this way the density of electrons in the conduction band increases, which leads, in the case of a reverse-biased p–n junction, to well-known avalanche breakdown. It is clear that this effect presents a limitation on the electric fields that can be used, and we shall illustrate this by band-to-band generation in a charge-coupled device. A simplified geometry of gates of a CCD was shown in Fig. 13. The electric fields are highest between neighboring gates and lowest beneath the middle of the gates. To achieve ultimate transfer speed, one wants to have the lowest electric field F_l larger than the field at which the drift velocity saturates. This is typically a field of 10^4 V/cm. On the other hand, the highest electric field F_h (between the gates) must not exceed the field at which impact ionization becomes important. This field is typically 10^5 V/cm. The lowest and highest electric field are connected by the approximate relation [64] for an oxide of thickness z_{ox}.

$$F_l \gtrsim 6.5(dz_{ox} F_h/L^2)$$

found by computer simulations. Typical values are $d \approx z_{ox} \approx 10^{-5}$ cm. Therefore, ultimate transfer speed is only possible for $L \lesssim 8 \times 10^{-5}$ cm, which gives $F_l = 10^4$ V/cm when $F_h = 10^5$ V/cm. Charge-coupled devices with longer gate lengths are slower. However, this is only valid for surface-channel CCDs, and for buried-channel CCDs one has to use different relations.

IV. GENERAL CONCLUSIONS

The low-field mobility of inversion-layer carriers is typically lower and less temperature-dependent than the mobility in bulk material. The lower mobility is caused by the additional scattering mechanisms that occur near the surface. These are mainly surface-roughness scattering and scattering by interface point charges. The smaller temperature dependence arises from the fact that surface-roughness scattering is only weakly dependent on energy over wide limits and also from the steplike density of states. Since almost all scattering mechanisms depend sensitively on the final density of states, the change in this quantity upon going from the three-dimensional bulk semiconductors, where it increases as $\sqrt{E}$, to the quasi-two-dimensional inversion layer, where it is independent of energy, is dramatic.

Where the electron temperature is sufficiently high, calculations of the transport of electrons in inversion layers in silicon yield very good agreement with the experimental data. The dominant intervalley phonon scattering is adequately accounted for by utilizing the bulk values of the coupling constants and taking account of the quasi-two-dimensional nature of the carriers. Repopulation of the various subbands is found to occur at moderate electric fields and the saturated drift velocity approaches that of bulk material.

At low lattice and electron temperatures, however, the situation is not as clear. Although values of the warm electron β can be calculated to agree with some values of that measured, the orders of magnitude variation in this measured quantity mirrors the problem of inversion-layer transport in general. That is, the exact nature of the scattering centers is just not well enough known in the degenerate inversion layer. As a consequence, the warm electron transport and, also, the magnetotransport of these layers provide valuable additional information for sorting out the transport properties.

A second area of future concern is the actual device configuration. To date, high field transport considerations have neglected to include inhomogeneity of the inversion layer itself, due to debiasing of the gate potential by the source–drain potential. To adequately model the quasi-two-dimensional effects on device performance, these inhomogeneities must be considered.

The possibility of extremely high electric fields and at the same time high gradients in the concentration of hot electrons makes it necessary to include the diffusion current and electronic heat conduction in the calculation of the electron temperature in devices. It follows that the mobility cannot be regarded as a single-valued function of the electric field. In these cases, approximations like $\mu = \mu_0/[1 + (E/E_C)^2]^{1/2}$ are therefore wrong. In a simplified treatment, E_C becomes a function of the concentration gradient as well, and this treatment should be still simple enough to be included in very simplified device models [67]. The electronic heat conduction also broadens the ranges of hot electrons in cases of inhomogeneous electric fields.

REFERENCES

1. F. Stern, *Crit. Rev. Sol. State Sci.* **4**, 499 (1974).
2. B. Hoeneisen and C. A. Mead, *Solid-State Electron.* **15**, 819 (1972).
3. F. Fang and A. B. Fowler, *J. Appl. Phys.* **41**, 1825 (1970).
4. D. K. Ferry, *Phys. Rev. B* **14**, 5364 (1976).
5. F. Stern and W. E. Howard, *Phys. Rev.* **163**, 816 (1967).
6. M. Abramowitz and I. A. Stegun, "Handbook of Mathematical Functions." National Bureau of Standards, Washington, D.C., 1964.
7. C. Jacoboni, C. Canali, G. Ottaviani, and A. Alberigi-Quaranta, *Solid-State Electron.* **20**, 77 (1977).
8. K. Hess, *Solid-State Electron.* **21**, 123 (1978).
9. A. Many, "Semiconductor Surfaces." North-Holland Publ., Amsterdam, 1965.
10. D. R. Frankl, "Electrical Properties of Semiconductor Surfaces." Pergamon, Oxford, 1967.
11. R. F. Greene, *Crit. Rev. Solid State Sci.* **4**, 455 (1974).

12. S. R. Hofstein and F. R. Heiman, *Proc. IEEE* **51**, 1190 (1963).
13. F. F. Fang and A. B. Fowler, *Phys. Rev.* **169**, 619 (1968).
14. H. Ezawa, S. Kawaji, and K. Nakamura, *Surf. Sci.* **27**, 218 (1971).
15. H. Ezawa, S. Kawaji, and K. Nakamura, *Jpn. J. Appl. Phys.* **13**, 126 (1974).
16. H. Ezawa, S. Kawaji, T. Kuroda, and K. Nakamura, *Surf. Sci.* **24**, 659 (1971).
17. C. T. Sah, T. H. Ning, and L. L. Tschopp, *Surf. Sci.* **32**, 56 (1972).
18. E. D. Siggia and P. C. Kwok, *Phys. Rev. B* **2**, 1024 (1972).
19. Y. C. Cheng, *J. Appl. Phys.* **44**, 2425 (1973).
20. F. Berg, *Solid-State Electron.* **13**, 903 (1970).
21. S. Kawaji and Y. Kawaguchi, *J. Phys. Soc. Jpn. Suppl.* **21** 336 (1966).
22. J. L. Rutledge and W. E. Armstrong, *Solid-State Electron.* **15**, 215 (1966).
23. K. Hess and C. T. Sah, *Phys. Rev. B* **10**, 3375 (1974); *J. Appl. Phys.* **45**, 1254 (1974).
24. C. M. Krowne and J. W. Holm-Kennedy, *Surf. Sci.* **46**, 197 (1974).
25. D. K. Ferry, *Surf. Sci.* **57**, 218 (1976).
26. J. M. Ziman, "Electrons and Phonons." Oxford Univ. Press, London and New York, 1960.
27. Y. C. Cheng and E. A. Sullivan, *J. Appl Phys.* **44**, 3619 (1973).
28. J. Mertsching and H. J. Fischbeck, *Phys. Status Solidi* **41**, 45 (1970).
29. Y. C. Cheng and E. A. Sullivan, *Surf. Sci.* **34**, 717 (1973).
30. A. Hartstein, T. H. Ning, and A. B. Fowler, *Surf. Sci.* **58**, 178 (1976).
31. J. W. Harrison and J. R. Hauser, *Phys. Rev. B* **13**, 5347 (1976).
32. D. K. Ferry, *Phys. Rev. B* **17**, 912 (1978).
33. M. A. Littlejohn, J. R. Hauser, T. H. Glisson, D. K. Ferry, and J. W. Harrison, *Solid-State Electron.* **21**, 107 (1978).
34. B. T. Moore and D. K. Ferry, *J. Appl. Phys.* **51**, 2603 (1980).
35. K. Hess, *Appl. Phys. Lett.* **35**, 484 (1979).
36. F. Stern *Phys. Rev. Letters* **44**, 1469 (1980).
37. K. Hess, R. Merlin, and N. V. Klein (to be published).
38. K. Hess and P. Vogl, *Solid-State Commun.* **30**, 807 (1979).
39. K. Hess, T. Englert, T. Neugebauer, G. Landwehr, and G. Dorda, *Phys. Rev. B* **16**, 3652 (1977).
40. W. A. Harrison, *Phys. Rev.* **104**, 1281 (1956).
41. F. Seitz, *Phys. Rev.* **73**, 549 (1948).
42. H. J. G. Meyer, *Phys. Rev.* **112**, 298 (1958).
43. E. Conwell, "High Field Transport in Semiconductors." Academic Press, New York, 1967.
44. M. Lax and J. J. Hopfield, *Phys. Rev.* **124**, 115 (1961).
45. H. W. Streitwolf, *Phys. Status Solidi* **37**, K47 (1970).
46. M. Lax and J. L. Birman, *Phys. Status Solidi (b)* **49**, K153 (1970).
47. D. K. Ferry, *Phys. Rev. B* **14**, 1605 (1976).
48. A. Fowler, *in* "Handbook of Semiconductors" (W. Paul, ed.), Vol. I. North-Holland Publ., Amsterdam (in press).
49. S. Q. Wang and G. D. Mahan, *Phys. Rev. B* **6**, 4517 (1972).
50. D. L. Long, *Phys. Rev.* **120**, 2024 (1961).
51. P. Norton, T. Braggins, and H. Levinstein, *Phys. Rev. B* **8**, 5632 (1973).
52. J. C. Portal, L. Eaves, S. Askenazy, and R. A. Stradling, *Proc. Int. Conf. Phys. Semicond., Stuttgart* (M. H. Pilkuhn, ed.) p. 259. Teubner, Stuttgart, (1974).
53. D. K. Ferry, *Solid-State Electron.* **21**, 115 (1978).
54. T. Sato, Y. Takeishi, H. Hara, and Y. Okamoto, *Phys. Rev. B* **4**, 1950 (1970).

55. A. G. Sabnis and J. T. Clemens, *Proc. Int. Electron Devices Meeting* p. 18. IEEE Press, New York, 1979.
56. N. O. Gram, *Phys. Lett. A* **38**, 235 (1972).
57. Y. Katayama, I. Yoshida, N. Kotera, and K. F. Komatsubara, *Appl. Phys. Lett.* **20**, 31 (1972).
58. K. Hess, A. Neugroschel, C. C. Shive, and C. T. Sah, *J. Appl. Phys.* **46**, 1721 (1975).
59. J. A. Cooper, Jr., and D. F. Nelson, *IEEE Trans. Electron Devices* **ED-27**, 2179 (1980).
60. P. Das and D. K. Ferry, *Solid-State Electron.* **19**, 851 (1976).
61. D. K. Ferry and P. Das, *Solid-State Electron.* **20**, 355 (1977).
62. S. A. Abbas and R. C. Dockerty, *Proc. Int. Electron Devices Meeting* p. 35. IEEE Press, New York, 1975.
63. T. H. Ning, *IEEE Trans. Electron Devices* **ED-25**, 1348 (1978).
64. L. Forbes, E. Sun, R. Alders, and J. Moll, *IEEE Trans. Electron Devices* **ED-26**, 1816 (1979).
65. K. Hess and H. Schichijo, *IEEE Trans. Electron Devices* **ED-27**, 503 (1980).
66. K. Hess and C. T. Sah, *IEEE Trans. Electron Devices* **ED-25**, 1399 (1978).
67. R. Stratton, *Phys. Rev.* **126**, 2002 (1962).
68. R. M. Kolbas *et al.*, *Solid State Commun.* **31**, 1003 (1979).

Chapter **4**

Impact of Submicron Technology on Microwave and Millimeter-Wave Devices†

P. A. BLAKEY, J. R. EAST, AND G. I. HADDAD

Electron Physics Laboratory
Department of Electrical and Computer Engineering
The University of Michigan
Ann Arbor, Michigan

† Research partially sponsored by the Air Force Office of Scientific Research, Air Force Systems Command, USAF, under Grant No. AFOSR-76-2939. The United States government is authorized to reproduce and distribute reprints for governmental purposes notwithstanding any copyright notation hereon.

105

I. INTRODUCTION

In the early years of semiconductor device fabrication the available technologies were largely limited to alloying and diffusion, coupled with conventional photoresist and chemical etch techniques. The available spatial resolution was limited to characteristic dimensions measured in microns. The introduction of new technologies such as ion implantation, molecular-beam epitaxy, and electron-beam lithography, coupled with continuous improvements in other areas, such as contact technology, has drastically improved the available resolution. Devices with submicron dimensions are now production items, while in research laboratories resolution is virtually down to the atomic monolayer level in one dimension, with linewidths of a few tens of lattice constants available in the other dimensions [1–3].

These advances in submicron semiconductor device technology are being exploited in several areas of application. First, as device dimensions are reduced, it becomes possible to incorporate more individual devices on a given area of semiconductor so the degree of integration per chip increases. Second, as device dimensions shrink there is a reduction in the time required for a change of electronic state, so semiconductor logic elements become faster. Third, as dimensions are reduced and transit times decrease, available frequencies of operation of active devices increase. This applies both to devices that are *limited* by transit-time effects and those that *exploit* transit-time effects. The purpose of this chapter is to review the impact of submicron technology on the last of these areas, i.e., on microwave and millimeter-wave semiconductor devices.

The approach adopted in this chapter is to concentrate on the possibilities for exploiting submicron technology in practical microwave and millimeter-wave devices. As far as possible in the available space, the underlying reasons for obtaining (or at least predicting) improved performance from devices fabricated using submicron technology are explained and experimental results obtained to date are summarized. The actual existence of submicron technology is accepted without discussion of the details of the techniques involved.

While many of the underlying principles may be simply explained in terms of the familiar ''drift plus diffusion'' model of carrier transport, which is widely applied to larger devices, the drift plus diffusion formulation is often inappropriate for detailed analysis of submicron semiconductor devices. Unless additional physics is appropriately included, quantitative results and scaling arguments will be, at best, approximate. The additional physics required is not yet widely appreciated, and the significant existing body of work that is relevant to submicron device modeling is not particularly well known. The next section, therefore, provides a brief review of the hierarchy of theories applicable to charge transport in semiconductors. In subsequent sections, which deal with particular microwave and millimeter-wave devices, underlying principles will in many cases be described in terms of the familiar drift plus diffusion model. However, throughout the chapter, it should be understood that this will not always be appropriate for detailed analysis.

The various microwave and millimeter-wave devices to be considered are divided into three categories, each of which is described in a separate section. They are as follows:

(1) Section III: Inherently multiterminal, multidimensional devices, such as various transistor structures;

(2) Section IV: Essentially one-dimensional, two-terminal, negative-resistance devices, such as IMPATTs and transferred-electron devices;

(3) Section V: New devices, such as superlattice and real-space-transfer devices, which have been made possible by submicron technology.

Each of these sections develops along similar patterns. An explanation of basic principles is followed by a discussion of the present and future impact of submicron technology and, where appropriate, a summary of experimental results. Section VI contains a brief discussion and conclusions.

II. CHARGE TRANSPORT IN SEMICONDUCTORS

In this section the various approaches that may be applied to semiconductor device modeling are reviewed and the approximations inherent in each are discussed. More details may be found elsewhere [4–6]. Our intent is to make clear the nature of the simplifications used elsewhere in the chapter and to introduce terminology to be used in subsequent sections.

A. The Quasi-Free-Particle Approximation

A hierarchical structure for the theory of charge transport in semiconductors is shown in Fig. 1. The most convenient starting point for discussing this is the quasi-free-particle (QFP) approximation, based on the following assumptions:

(1)   The quantum-mechanical nature of electron transport in a perfect periodic (i.e., lattice) potential is contained in periodic energy versus wave vector relationships applicable to point "particles" (band theory). It is assumed that all the other factors that affect the electrons may be treated as perturbations of the basic band picture.

(2)   The behavior of the "particles" may be described statistically using the classical Boltzmann theory of dilute gases. [This builds on the

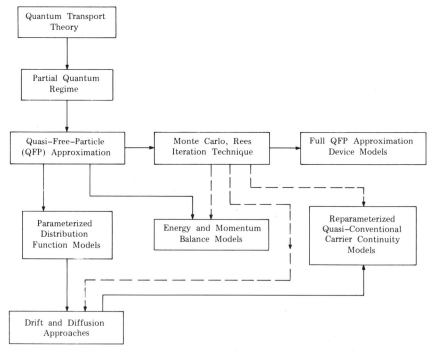

Fig. 1.   A hierarchy of models for charge transport in semiconductors.

assumptions that scattering processes are binary (only two elements per collision) and instantaneous.]

(3) The scattering rates associated with processes that constitute departures from the ideal lattice situation can be calculated from first-order, time-dependent, perturbation theory.

Each of items (1)–(3) is an approximation that breaks down under certain circumstances (see, for example, Barker [7]). However, for characteristic dimensions down to approximately 0.1 μm and for applied electric fields that are not too high, the description provided by the QFP approximation is generally good. Once the band structure and scattering rates have been calculated, the Boltzmann equation can be solved without further assumptions, using either a statistical (Monte Carlo) simulation [8,9] or the iterative technique developed by Rees [10]. Over the past few years Monte Carlo techniques have become popular as a means of studying and characterizing charge transport in semiconductors. They are simple to program but, although they have been applied to device simulations [11–14], they have proved to be prohibitively expensive for routine device analysis.

B. The Sub-QFP Regime

The sub-QFP regime is reached when additional approximations are made in order to arrive at affordable device models. One possibility is to make an assumption regarding the *form* of the distribution function to within some (usually two) adjustable parameters. The most usual assumptions are of a displaced Maxwellian distribution [4,15] or of a truncated Legendre polynomial expansion [4,16]. The method of moments can be applied to build the requirements of energy and momentum conservation into the model (see, for example, Blotekjaer [15]) and the collision terms in the Boltzmann equation can be evaluated [17,18]. This approach leads to tractable device models which have been applied to transferred-electron devices [19,20] and are currently being applied to avalanche diodes [21].

Another approach, which is conceptually a further approximation, is to neglect the *structure* of the distribution in k space and to conserve only *mean* energy and momentum. It then becomes possible to write simple energy and momentum *balance* relationships using phenomenologically defined energy and momentum relaxation times that are functions only of mean energy. This approach, originally established many years ago [4,22], has recently been popularized by Shur [23]. Although it is an apparently crude procedure, it can be very successful, since, if Monte Carlo simulation is used to obtain the relaxation time data, no explicit assumption regarding the form of the distribution function need be made, but a realistic distribution function will be implicit in the model. The combination of Monte Carlo data and energy and momentum balance seems capable of excellent results at low cost [24–26].

The usual drift plus diffusion approach represents a further, rather drastic, approximation. It corresponds to assuming that for each electric field there is an equilibrium form of the carrier distributions and that the energy and momentum relaxation times are so short with respect to all other times of interest in device operation that the carrier distributions always have the equilibrium forms appropriate to the instantaneous value of the local electric field. Transport parameters (commonly drift velocities, diffusion coefficients, and ionization rates) are then also functions only of the instantaneous electric field. The breakdown of this approach in submicron semiconductor devices is due to energy and momentum relaxation times (and in some cases possibly intervalley scattering times) becoming significant with respect to other characteristic device times.

Another sub-QFP approach starts from the drift and diffusion approximation and is applicable to the regime where the drift plus diffusion approach is almost appropriate, but some additional effect (or possibly effects) must be included. In this situation it is sometimes possible to reformulate the carrier continuity equations to include a description of the additional effect(s). Additional transport parameters may be defined and carrier types may be subdivided (into type "A" electrons and type "B" electrons, for example) with each carrier type having its own continuity equation. The label "reparameterized quasi-conventional carrier-continuity model" is sometimes applied to this approach. It has been used to include carrier–carrier effects on the ionization process in IMPATT modeling [27], dead spaces in the ionization process [28], and, in a variety of attempts (reviewed by Blotekjaer [15]), dynamic effects in transferred-electron devices.

C. Partial and Full Quantum Regimes

Above the QFP approximation in Fig. 1 there is the "partial quantum regime." Here the QFP approximation is substantially valid, but additional quantum effects must be imposed on the basic framework. Examples are tunneling [29] (interband transitions induced by high electric fields), size quantization (quantization of the energy levels associated with carrier motion which is highly constrained in one dimension, e.g., in very narrow inversion layers in MOS devices [30–32]), and, in certain situations, carrier transport in heterostructures and superlattices. Finally, as devices become smaller still, one is forced into a full quantum treatment. Until recently this was the realm of the theoretical physicist and only recently has the immense importance to technologically feasible devices started to be appreciated [6,7].

D. Discussion

Within the hierarchy of Fig. 1 only the drift plus diffusion approximation is familiar to most device engineers. This is not surprising since it is conceptually simple and, until recently, was adequate for most technologically feasible devices. Also, even this "simple" model results in three coupled, nonlinear, partial differential equations whose solution requires either further simplifying approximations or the use of numerical methods. (In fact, numerical methods of solution even for this basic model were largely developed in the last ten to fifteen years, and are generally quite complicated and expensive if more than one spatial dimension is included.)

There are, however, few fundamental difficulties in implementing sub-QFP approximation models that are significantly better than the drift plus diffusion model. When energy and momentum transport are taken into account using parameterized distribution functions, the resulting set of equations has distinct similarities to familiar fluid-flow equations for which numerical solution techniques are well established. The energy and momentum balance approach has already been noted as particularly cost-effective when used in conjunction with Monte Carlo generated parameters.

In the full QFP approximation there are also no difficulties *in principle* in implementing device models, but cost and storage requirements are prohibitive. This situation may change in the future, but at the moment the main role of Monte Carlo simulation is to provide parameters for sub-QFP approximation device models. This is indicated in Fig. 1.

Difficulties increase in the quantum regime. The partial quantum regime (tunneling, size quantization, etc.) can be handled by ad hoc modification of the QFP and sub-QFP models, but it is not yet clear how device models will be implemented in the full quantum regime. Contacts (boundary conditions) will present particular problems. It will also be difficult to avoid thinking in terms of the QFP approximation and to adjust to the idea of specifically seeking to exploit quantum effects in the device structures that are becoming technologically feasible.

III. TRANSISTOR STRUCTURES

The first class of submicron devices to be considered is that of multiterminal, multidimensional, transistor structures. The well-known transistor structures are bipolar junction transistors (BJTs), junction-gate field-effect transistors (JFETs), and insulated-gate field-effect transistors (IGFETs). Each type has performance characteristics that suit it to particular applications, both linear and digital and in discrete and integrated forms. Submicron technology is important to all these structures, but when the discussion is limited to microwave and millimeter-wave devices, JFETs (and, in particular, GaAs MESFETs) are currently much more important than the other structures. In this section the discussion is, therefore, centered on MESFETs. Bipolar transistors are important mainly at the low end of the microwave frequency range, below approximately 6 GHz. A review of bipolar transistors at microwave frequencies is given by Allison [33]. Details of work on IGFETs at microwave frequencies may also be found in the literature [34,35].

The organization of this section is as follows. The basic principles of FET operation are presented in Section III.A and a very simple analytical model is used to establish the basic advantages of FETs with submicron dimensions. An improved quasi-two-dimensional model which gives information regarding the effects of space-charge, nonlinear velocity–field characteristics, and diffusion is presented in Section III.B. Nonequilibrium hot-carrier effects are reviewed in Section III.C, variations on the basic FET are outlined in Section III.D, and the present state of the art in microwave transistors is discussed in Section III.E.

A. FET Principles

Many of the ideas underlying FET operation were presented by Shockley [36]. A general review of microwave FETs was given by Pucel *et al.* [37], and recent reviews of microwave FET work were given by Liechti [38] and DiLorenzo and Wisseman [39].

The very basic principles of FETs are rather simple. A slab of semiconductor with ohmic contacts (drain and source) at each end (Fig. 2a) behaves like a resistance, at least for fields low enough for the mobility to be constant. If a Schottky contact (gate) is placed on the slab, as shown in Fig. 2b, a depletion region will be formed under the gate. This depletion region will decrease the cross-sectional area and thereby increase the drain–source resistance. The net result is that an input voltage on the gate

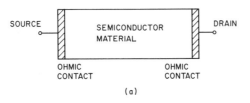

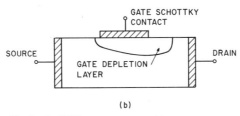

Fig. 2.   The basic FET structure: (a) without gate; (b) with gate.

controls the drain–source output current (the "linear" region of Fig. 3), giving a voltage-controlled resistance.

In practice the FET is not normally biased in this linear region of its characteristic, but at higher drain–source voltages. In this case the fields under the gate become high enough for velocity saturation to occur. Also, associated with field gradients (through Poisson's equation), there will be carrier concentrations significantly different from the background doping density. This leads to accumulation/depletion layers (charge dipoles) which tend to be located under the gate around the drain end. A considerable amount of the drain–source voltage can be dropped across such a dipole. The combined effects of the velocity saturation and the voltage drop across a charge dipole cause the drain–source current to very nearly saturate as a function of drain–source voltage (Fig. 3). When operated in the saturated regime, the most important device parameter is the transconductance, which is given by

$$g_m = \partial I_{ds}/\partial V_{gs},$$

where I_{ds} is the drain–source current, V_{gs} the gate–source voltage, and g_m the transconductance.

The structure just discussed is a MESFET (metal–semiconductor FET), the metal referring to the Schottky contact. It is, of course, possible to use a p–n junction rather than a Schottky contact. This can give a better-quality junction and is done for lower-frequency devices, but for small-dimension microwave devices it is easier to fabricate MESFETs. The ratio of transconductance to gate capacitance provides a useful guide to the effect of geometry on high-frequency FET operation. In general,

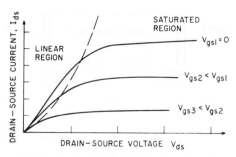

Fig. 3.   Basic FET electrical characteristics.

the higher the ratio, the better the high-frequency performance is expected to be. The reasons for this can be seen by considering the factors affecting the duration of the transient Δt associated with charging the gate capacitance C_g following a change ΔV_{gs} in the gate voltage. If ΔI_c is the mean current available for charging the gate capacitance, then

$$\Delta t \approx C_g \, \Delta V_{gs}/\Delta I_c.$$

It is reasonable to assume that ΔI_c is approximately proportional to ΔI_{ds}, the change in I_{ds} associated with ΔV_{gs}. In this case

$$\Delta t \propto C_g/g_m.$$

The shorter the duration of the transient, the better the high-frequency performance, so g_m/C_g can be used as a figure of merit for high-frequency operation. The basic advantages of submicron geometry FETs can be explained on the basis of a simple idealized model due to Pucel et al. [37]. This will be outlined here and the results applied to show the advantages of submicron geometry for high-frequency FET operation. The assumed geometry and associated notation (which is the same as used in [37]) are shown in Fig. 4. The idealized two-line velocity-field characteristic used consists of a constant mobility for fields up to E_s and a saturated velocity

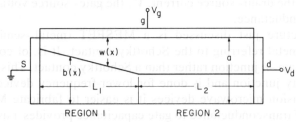

Fig. 4.   Device geometry for the analytic FET model.

v_s equal to μE_s for higher fields. The model does not include diffusion. In Fig. 4 the conducting channel is divided into two regions. In region 1 between $x = 0$ and $x = L_1$ the mobility is constant, and in region 2 between $x = L_1$ and $x = L_2$ the velocity is constant. The distance L_1 and the potential are found using current continuity and the total voltage drop between drain and source. In region 1 the gradual channel approximation [36] is used to find the conducting channel height. In region 2 the conducting channel height is taken as constant and equal to the height at $x = L_1$. In region 1 the current is (assuming constant mobility)

$$I_d = Zb(x)\sigma E(x), \tag{1}$$

where I_d is the drain current, Z the channel width, $b(x) = a - w(x)$, and σ is the conductivity. In region 2 the current for constant velocity is

$$I_d = Zb(x = L_1)\sigma E_s. \tag{2}$$

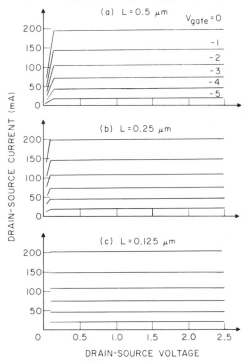

Fig. 5. Current–voltage characteristics for different gate lengths: (a) 0.5 μm; (b) 0.25 μm; (c) 0.125 μm.

Equation (1) can be integrated to find the potential drop across region 1 and a power series expansion of Laplace's equation can be used to find the potential across region 2. The resulting equations have been incorporated into a computer program which can be used to calculate the dc and small-signal parameters associated with the model with device geometry, doping, drain voltage, and gate voltage as input parameters. These results can be used to establish certain features of FET operation in the limit of small geometry.

To illustrate this, three small-geometry FETs with gate lengths of 0.5, 0.25, and 0.125 μm will be considered. Other device parameters are taken as typical of X-band FETs: channel height = 0.3 μm, channel width = 500 μm, and channel doping = $10^{17}/cm^3$. Figure 5 shows the drain current versus drain–source voltage for various gate voltages and suggests that the variation in the gate length has only a small effect on the output current–voltage characteristics. The drain current for the drain–source voltages above the linear region is dominated by the parameters in Eq. (2). For reasonable bias voltages the electric field in the channel reaches the saturation field over a short distance (region 1), and most of the poten-

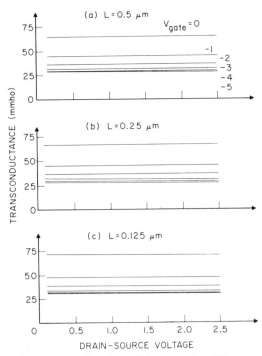

Fig. 6. Transconductance as a function of bias point for different gate lengths: (a) 0.5 μm; (b) 0.25 μm; (c) 0.125 μm.

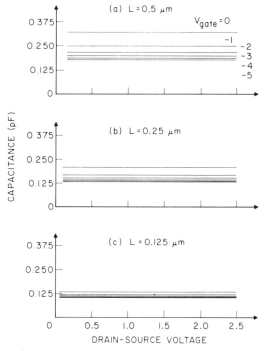

Fig. 7. Gate capacitance as a function of bias point for different gate lengths: (a) 0.5 μm; (b) 0.25 μm; (c) 0.125 μm.

tial drop occurs across region 2. When the length of the saturated velocity region is changed by changing the gate length, the electric field in the saturated velocity region changes but does not greatly affect the $b(L_1)$ term in Eq. (2) so the current remains approximately constant.

The transconductance and capacitance for the three structures are shown in Figs. 6 and 7. Since the drain current is approximately the same at a given bias point in each of the three transistors, the transconductance is also approximately constant (there is, in fact, a slight increase in transconductance with decreasing gate length). The gate capacitance depends on the area of the gate electrode and the extension of the depletion layer to the surface at the edge of the gate. It is here that the improvement in device performance with reduced dimensions is most apparent, because, except for the edge capacitance, there is a linear decrease in capacitance with length. (The capacitance can also be decreased by reducing the channel width Z, but this would reduce the drain current and transconductance by a corresponding factor and not improve the C_g/g_m ratio.) Thus it may be concluded that reducing the gate length is the most obvious way of improving the high-frequency performance of FETs. It is the ability to fabricate FETs with submicron gate lengths that has led to the excellent microwave performance from GaAs MESFETs.

B. A Quasi-Two-Dimensional FET Model

The model used in the previous section is useful for establishing certain basic features of FET operation, but even within the drift and diffusion approximation it neglects space-charge effects, diffusion, and realistic velocity–field characteristics. Traditionally, these have been investigated using a full two-dimensional model employing the numerical techniques of finite differences [40–43] or finite elements [44,45]. However, virtually all the features predicted by these rather complicated and expensive programs may be reproduced using a much simpler quasi-two-dimensional model [46]. Results from this quasi-two-dimensional model will be presented in this section to provide additional insight into FET operation.

The geometry used in the model is shown in Fig. 8. Voltage, field, velocity, and diffusion coefficients are assumed to be functions of x but not of y, and the depletion width $D(x)$ is a function of voltage along the channel. For a steady-state solution the electron current density J is a function of x, but the electron current JB is not. An analytic expression is used for the gate depletion layer width (this must be more sophisticated than the normal gradual channel approximation [36] to obtain accurate results [46]). Poisson's equation and the current conservation equation can then be solved numerically to find the steady-state electric field and electron distribution in the channel of the FET. Although the model may appear crude, comparison with full two-dimensional simulations demonstrates agreement within 10% for potential and charge distributions within the channel [46]. Given the uncertainties in appropriate material parameters (and also in the validity of the underlying drift and diffusion treatment of electron transport), the quasi-two-dimensional model appears much more cost-effective than "full" two-dimensional simulations as a means of investigating FET properties calculated in the drift and diffusion approximation.

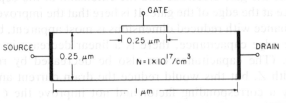

Fig. 8. Geometry and dimensions for the quasi-two-dimensional FET model.

Typical results for the GaAs MESFET structure of Fig. 8 are shown in Figs. 9 and 10. These use the following velocity–field $v(E)$ and diffusion coefficient–field $D(E)$ characteristics:

$$v(E) = \frac{5000E + 8 \times 10^6 \, (E/3.5 \times 10^3)^4}{1 + (E/3.5 \times 10^3)^4} \quad \text{cm/sec}$$

and

$$D(E) = \frac{130 + 20(E/4.5 \times 10^3)^4}{1 + (E/4.5 \times 10^3)^4} \quad \text{cm}^2/\text{sec}.$$

An important feature of the results is that a charge dipole forms for electric fields in the channel higher than the field for peak velocity. It extends from the narrowest part of the channel, near the drain end of the gate, to the drain edge of the gate depletion layer. The high electric field in the charge dipole reduces the field in the rest of the channel, keeping the velocity high. Most of the potential drop and charge imbalance occur across and within the charge dipole. Velocities outside the charge dipole are in the approximately constant mobility region. Diffusion spreads out the edges of the charge dipole, but the dipole still gives rise to large field gradients (the field rises from values less than 3 kV/cm to values in the range of 50 to 100 kV/cm over a very short distance). Under these

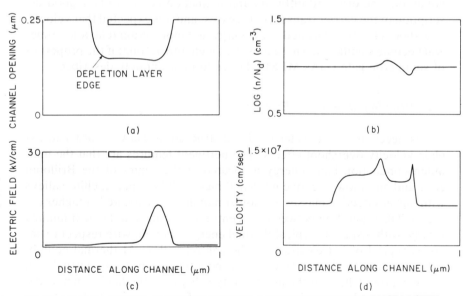

Fig. 9. Results for $V_{gs} = 0$ V and $V_{ds} = 0.4$ V: (a) channel depletion; (b) location and extent of charge dipole; (c) electric field versus distance; (d) electron velocity versus distance.

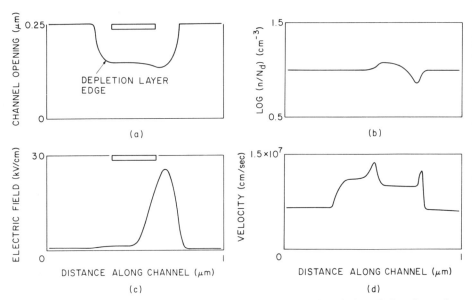

Fig. 10. Results for $V_{gs} = 0$ V and $V_{ds} = 0.8$ V: (a) channel depletion; (b) location and extent of charge dipole; (c) electric field versus distance; (d) electron velocity versus distance.

conditions the drift and diffusion approximation ceases to be a good description of the carrier transport processes, and it is expected that velocity overshoot effects associated with energy and momentum relaxation times will become significant. These must be taken into account if the properties of submicron gate GaAs MESFETs are to be accurately described.

C. Velocity Overshoot Effects

It is necessary to consider the band structure of GaAs in order to explain velocity overshoot effects. The pertinent features are that the minimum conduction band energy is located at the center of the Brillouin zone and has a low effective mass and high mobility, and satellite valleys with higher effective masses and lower mobilities are located at higher energies. The separation between the central minimum and the next minima is large with respect to typical thermal energies but not with respect to energies typical of hot electrons in the central valley. At thermal equilibrium, and for low electric fields, the conduction electrons are virtually all in the central valley, so the net mobility is high. As the field is increased, significant numbers of electrons acquire sufficient energy to transfer to the satellite valley and the mean velocity decreases. Finally, at very high fields, almost all the electrons are in the satellite valleys and the mean velocity saturates. This explains the form of the static velocity–field characteristic used in the quasi-two-dimensional model.

In Section II it was explained that the use of a static velocity–field characteristic ceases to be a good approximation when the field is changing rapidly. This is because of the neglect of energy and momentum relaxation times and of nonequivalent intervalley scattering times. The net result of this in GaAs MESFET operation is that as an electron enters the dipole region, it tends to be accelerated by the field to velocities significantly greater than the static peak velocity. The electron continues to accelerate for the time (and distance) it takes for scattering into the satellite valley. The details of the overshoot depend on the field step and the relative magnitudes of the energy and momentum relaxation times [47–49]. For the field gradients encountered in FETs, electrons may travel a few tenths of a micron before the scattering occurs. This is comparable with the length of the charge dipole. Since the mean electron velocity is higher than that predicted on the basis of the drift and diffusion approach, both the electron concentration within the dipole and the potential drop across it are lower. This should increase the drain current at a given bias point thereby improving g_m and the high-frequency performance.

Further modeling to take into account the relevant physics is needed to investigate the effects of velocity overshoot in detail. A Monte Carlo approach was adopted by Warriner [14] and an energy and momentum balance approach by Carnez et al. [26]. Both of these analyses seem to confirm the qualitative arguments regarding the steady-state properties. There remains a real need for a dynamic large-signal FET simulation that incorporates the velocity overshoot effects.

D. Permeable-Base and Ballistic Transistors

Two devices which have recently attracted interest for high-frequency operation are the permeable-base transistor (PBT) [50,51] and the ballistic transistor [52]. These may be regarded as variations on the previously discussed FET concepts.

An experimentally fabricated PBT device [50] is shown schematically in Fig. 11. The gate consists of a grid of tungsten fingers formed by laser holography, x-ray lithography, and lift-off. The finger thickness is 200 Å and the repeat spacing is 3200 Å. Epitaxially grown single-crystal GaAs surrounds the gate. The source and drain contacts are on the top and bottom of the structure, respectively, so the source–drain current flows

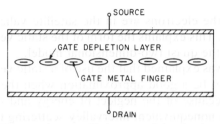

Fig. 11. A permeable-base transistor structure.

between the gate fingers and is controlled by the depletion region surrounding each finger. The depletion region is in turn controlled by the gate voltage. A two-dimensional simulation of the PBT in the drift and diffusion approximation suggested a maximum frequency of oscillation of 300 GHz [51]. However, dynamic velocity overshoot effects would have to be included in any accurate analysis of transport in the gate region (especially at high frequencies), and parasitic reactances, which were neglected in the simulation, are acknowledged to be high [50,51]. In fact it is not obvious on the basis of simple transconductance/gate capacitance arguments that the high-frequency performance of the PBT will be significantly better than that of other forms of FETs.

The ballistic transistor is essentially an FET with the velocity overshoot effect taken to the extreme of no scattering events [52]. In this limit the electron velocity depends on the potential not on the electric field. The device structure is similar to a conventional FET, but operation is at low temperatures to reduce phonon scattering. Because of the low probability of scattering, the electrons can reach a high velocity and so the current can be higher than in a comparable FET structure. This should improve the transconductance and hence high-frequency performance of the device.

The original analysis of the ballistic transistor contains a number of approximations and idealizations. It is not clear to what extent these contribute to an overoptimistic estimate of the performance that will be obtainable in practice. Maloney [53] pointed out that the assumed mean free path of 1 μm is too long; it is appropriate to low-energy electrons but a much shorter mean free path is expected for practical potentials such that the threshold for polar optical phonon emission is exceeded. Thus there is considerable uncertainty regarding what, if any, performance advantages will result from ballistic mode operation of FETs.

E. State of the Art Performance

The power, efficiency, noise, and high-frequency performance of conventional microwave GaAs MESFETs are still rapidly improving. Experi-

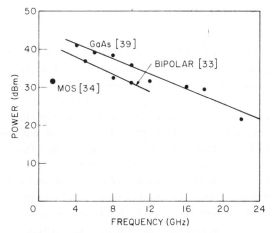

Fig. 12. State of the art for transistor power.

mental FETs are being fabricated with gate lengths in the range of 0.2 to 0.5 μm. The performance tends to be limited by factors such as chip parasitics and impedance-level and frequency-matching limitations in the external circuit. Some representative state of the art results are shown in Figs. 12 and 13 [33,34,39,54–57]. These are highly competitive with other semiconductor devices over the frequency range indicated. Since the FET also has other advantages (these are detailed by Liechti [38] and include self-ballasting (avoidance of thermal runaway), high reverse isolation, and radiation resistance), the GaAs MESFET seems to be very suitable for many applications up to at least 30 GHz and eventually possibly to significantly higher frequencies.

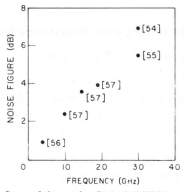

Fig. 13. State of the art for GaAs MESFET amplifier noise.

IV. TWO-TERMINAL DEVICES

Long before GaAs MESFETs proved to have such excellent performance characteristics at microwave frequencies, a variety of two-terminal
negative-resistance devices were developed. These devices included
IMPATTs, transferred-electron devices (Gunn diodes), and BARITTs.
For many applications, at frequencies below approximately 30 GHz, these
devices have been (or with high probability, soon will be) superseded
by GaAs MESFETs. However, at significantly higher frequencies and for
some high-power applications, two-terminal negative-resistance devices
will continue to be the most important solid-state sources. In addition, the
nonlinearity of passive two-terminal devices can be exploited. Active and
passive two-terminal devices useful at microwave and millimeter-wave
frequencies will be considered in this section, organized as follows:
Two-terminal negative-resistance devices are divided into two categories,
"pure" transit-time devices, in which carriers are injected by some mechanism into a drift region (e.g., IMPATTs, BARITTs, and TUNNETTs),
and transferred-electron devices, which have various modes employing
combinations of negative mobility, space-charge instability, and transit-
time effects. The basic principles of operation of each category are reviewed and it is concluded that submicron technology will impact mainly
on the pure transit-time devices. The existing and proposed ways of exploiting submicron technology are split into doping control and material
control (heterostructure) approaches and each is discussed separately.
The use of small area diodes as passive nonlinear circuit elements useful
for frequency multiplication, mixing, and detection at millimeter-wave
frequencies is then briefly reviewed, and the factors which limit the areas
of both active and passive diodes at millimeter-wave frequencies are considered. Finally, the state of the art for power generation as a function of
frequency is discussed.

A. Basic Principles of Negative-Resistance Devices

1. Principles of Pure Transit-Time Devices

Pure transit-time devices are based on the injection of charge into a
region of a semiconductor device followed by drift across that region.
Each mechanism (injection and drift) introduces phase delays for the particle current components of the total current, and if the net phasing is correct, a negative resistance results [58].

The basic features of the drift process are described by the Ramo–
Shockley theorem [59,60] which gives the result that for the situation
shown in Fig. 14a the current induced in the external circuit by moving

charge is

$$J_{\text{ind}} = \frac{1}{w} \int_0^w Q(x)v(x)\ dx.$$

Thus a sharp pulse of charge injected at $x = 0$, traveling to $x = w$ at a constant velocity v_s, will induce a rectangular current pulse as shown in Fig. 14b. It will be assumed that the charge is injected as the result of the presence of a sinusoidal terminal voltage and that injection occurs at a phase angle θ_{inj}. In terms of the notation of Fig. 14b, it is straightforward to show that the output power per unit area is given by

$$-J_{\text{dc}}V_{\text{rf}}\left(\frac{(1 - \cos\theta_{\text{d}})\cos\theta_{\text{inj}} + \sin\theta_{\text{inj}}\sin\theta_{\text{d}}}{\theta_{\text{d}}}\right)$$

and the efficiency by

$$-\frac{V_{\text{rf}}}{V_{\text{dc}}}\left(\frac{(1 - \cos\theta_{\text{d}})\cos\theta_{\text{inj}} + \sin\theta_{\text{inj}}\sin\theta_{\text{d}}}{\theta_{\text{d}}}\right),$$

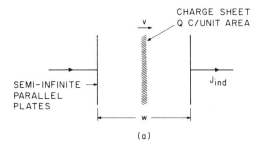

(a)

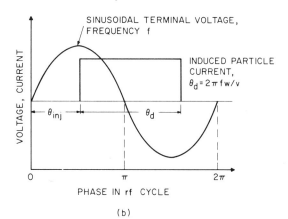

(b)

Fig. 14. Currents induced by moving charges: (a) physical situation; (b) induced current pulse.

where J_{dc} is the direct current density, V_{rf} the rf voltage, V_{dc} the dc voltage and θ_{d} the drift angle. These expressions describe the basic features of the dependence of the performance of transit-time devices on injection angle and drift angle. The model is, of course, highly simplified because in practice the injected charge pulse is not sharp and not all charge transport is at a constant velocity. It is clear from these expressions that the optimum drift angle, that which maximizes output power and efficiency, is a function of injection angle, as is the resulting output power and efficiency. Figure 15 is a plot of optimum drift angle and relative efficiency as functions of θ_{inj}. Clearly it is advantageous to have θ_{inj} as close as possible to $3\pi/2$ and to have a small θ_{d} (the class C limit), but in practice θ_{inj} is limited by the mechanisms available for charge injection. These will be considered next.

Three injection mechanisms have been exploited to date. These are injection over a potential barrier (the BARITT [61–65]), injection by field-induced, band-to-band tunneling (the TUNNETT [66–68]), and injection by avalanche multiplication (the IMPATT [69–73]). There is also a hybrid device where tunneling and avalanche multiplication exist simultaneously, which has been termed the MITATT [68]. Both injection over a potential barrier and tunneling are charge-injection mechanisms which are approximately in phase with the rf voltage and have a θ_{inj} of around $\pi/2$. Avalanche multiplication introduces a phase delay because the rate of charge "injection" depends on the quantity of charge present as well as the field level. Because of this, θ_{inj} is much closer to π. The MITATT will have θ_{inj} between that of the TUNNETT and the IMPATT, depending on how much of each injection mechanism is present. (All the injection

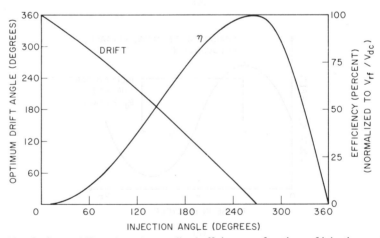

Fig. 15. Optimum drift angle and normalized efficiency as functions of injection angle.

mechanisms are in practice affected by the fields associated with previously injected charge. This tends to degrade the injection angle.) Although the IMPATT is predicted to have output powers and efficiencies significantly greater than the BARITT and the TUNNETT, because of the more optimum charge-injection phase angle, it will also be noisier; the BARITT and the TUNNETT are subject to injection-shot noise and velocity-fluctuation noise while the IMPATT has, in addition, multiplication noise and avalanche noise which tend to be much greater. Both the BARITT and the IMPATT give performances similar to those predicted by theory around X band (10 GHz) and there is no reason to believe that TUNNETTs would not also. However, with the competition from GaAs MESFETs at this frequency, many applications of two-terminal devices will be at significantly higher frequencies and it is therefore important to consider the high-frequency limitations of the various injection mechanisms [68,74,75].

The process of charge injection over a potential barrier is not particularly fast, since it involves diffusion-type processes. However, the important high-frequency performance-limiting mechanism in BARITTs appears to be the pulse spreading due to diffusion, which takes place at low fields in the drift region, immediately after charge injection. This seems to limit the usefulness of the BARITT in its active mode to frequencies below approximately 50 GHz. Tunneling is an extremely fast process; compared to other times of interest in millimeter-wave device operation, it is effectively instantaneous. Impact ionization is reasonably fast but has limitations associated with threshold energy requirements. On a simple picture this threshold energy would be taken as the bandgap energy; satisfying energy and momentum requirements simultaneously for realistic band structures gives a threshold energy closer to twice the bandgap energy [76–78]; taking into account the intracollisional field effect [79,80] may reduce the threshold somewhat. For a threshold energy of 2 eV and a field of 5×10^5 V/cm, a newly generated carrier must travel 0.04 μm ballistically to acquire the threshold energy. For an effective mass equal to the free electron mass, this corresponds to a minimum time delay of 0.95×10^{-13} sec. Looked at differently, in a device with a total potential of 5 V across its terminals (a typical value for 200 GHz IMPATTs), a given electron is unlikely to initiate more than two ionization events because of the threshold energy requirements. This limitation on multiplication reduces the particle current modulation depth, hence, output power and efficiency. It may be concluded that the impact ionization threshold energy requirement translates into a high-frequency performance-limiting mechanism (and also that energy and momentum conservation must be taken into account when modeling high-frequency IMPATTs [21]).

The dynamics of the ionization process in GaAs deserve special mention because they are still not adequately understood. Berenz *et al.* [81] showed, on the basis of experimental measurements, that there is a systematic discrepancy between calculated and measured noise and small-signal properties of GaAs IMPATTs. This discrepancy appears to be equivalent to an actual avalanche response time equal to approximately three times the calculated avalanche response time. Elta and Haddad [68] showed that phenomenologically tripling the response time in calculations of IMPATT performance leads to a rapid fall off in the predicted performance of GaAs IMPATTs as a function of frequency. There have been attempts to investigate the band structure dependence of the ionization process [82–84] but these have yet to provide a convincing explanation of the experimental results. It is also possible that other effects, such as recombination or carrier–carrier interactions [85], contribute significantly to the observed small-signal and noise performance.

The drift process is also not as ideal as assumed in the simple expressions. For various reasons (primarily ease of fabrication and optimization for maximum output power rather than maximum efficiency), devices are not operated with high fields throughout the drift region and velocities are not always "saturated." In fact, there is usually undepleted epitaxial material present in transit-time devices. Nonsaturation of the velocity of the generated charge and losses in undepleted epitaxial material are important performance-limiting mechanisms [86]. Also, it is not always appropriate to use the "static" velocity–field characteristic to estimate these losses at large signal levels and millimeter-wave frequencies. The question of epitaxial losses provides another doubt regarding the use of GaAs as a material for high-power millimeter-wave transit-time devices. Although traditionally thought of as low-loss epitaxial material, GaAs can become highly lossy at large signal levels as a result of electrons being "hung up" in the low-mobility satellite valleys. The losses can, in fact, be much worse than in Si under similar ac conditions. Germanium, with its siliconlike band structure but higher mobility, may prove to be a more useful material for millimeter-wave transit-time devices.

2. Principles of Transferred-Electron Devices

The mechanism whereby certain materials such as GaAs and InP have a negative differential mobility portion in their static velocity–field characteristic was explained in Section III.C. In this section the means of exploiting this mechanism to obtain microwave and millimeter-wave power from two-terminal devices will be outlined. The single most important point to note is that even if it were valid to assume a "static" velocity–field characteristic, a uniform piece of material could not be biased in the negative mobility portion of the characteristic. There is an inherent instability in the situation and any fluctuation leads to a rearrangement of free charge from the uniform case. Conceptually, there are two ways of exploiting the transferred-electron effect. A bar of uniform material may be biased in the negative mobility part of the characteristic to obtain a negative resistance upon application of an rf voltage. The incipient instability is "quenched" each cycle by driving the material into the positive mobility part of the velocity–field characteristic. This is termed the LSA mode. It is inherently large signal and capable of very high output powers but requires very uniform material. It is circuit controlled but has the drawback of requiring multifrequency circuit techniques. The alternative technique is to allow the instability to grow. A charge dipole accumulation/depletion layer or "domain" results, which can absorb a great deal of voltage; away from the domain the field drops, which tends to reduce the current. The domain is generally nucleated by a cathode doping notch and propagates to an anode in a transit time largely determined by the diode length, after which a new domain is nucleated and the process repeats. The oscillation is largely transit-time controlled. Output powers and efficiencies are generally low because the current and voltage modulation depths tend not to be very high. Depending on the diode doping, length, and operating frequency, a range of distinguishable modes may be predicted. Detailed discussions of the various modes of transferred-electron devices may be found elsewhere [87–89].

Transferred-electron devices have been quite widely used at microwave frequencies since they are simple to fabricate and relatively low noise. However, with the emergence of the GaAs MESFET the use of transferred-electron devices at microwave frequencies is likely to reduce and the higher-frequency (millimeter-wave) performance limitations must be examined.

It is inappropriate to use static velocity–field and diffusion coefficient–field models to study the high-frequency limitations because most of the important limiting mechanisms (energy and momentum relaxation times, intervalley scattering times, and details of the k-space structure) are ignored in the "static" limit. Instead, the physics must be studied in more detail. This has been done by Jones and Rees [90]. It seems that useful high-frequency operation of GaAs transferred-electron devices is limited by the fundamental physics of the material to frequencies not much higher than approximately 100 GHz; InP transferred-electron devices may have some high-frequency advantages as compared to GaAs devices [91]. It should be noted that the effects that determine the operation of transferred-electron devices are present in and may significantly affect the properties of the drift regions of pure transit-time devices made from GaAs and InP.

3. Discussion

Transferred-electron devices seem limited by inherent physics rather than by the dimensions of devices it is technologically feasible to build. This is in contrast to the pure transit-time devices where technological constraints are important. For the IMPATT the underlying physics spoils operation at high enough frequencies, as discussed earlier, but the impact of submicron technology is very important. With TUNNETTs the achievable performance appears largely determined by technology. Because of these considerations the remainder of the discussion concerning active devices is devoted to the impact of submicron technology on IMPATTs, MITATTs, and TUNNETTs.

B. Doping Control

In principle, any transit-time device can have its performance optimized, by tailoring the profile, for operation at given operating conditions. In practice, the amount of tailoring that is done is limited for two reasons: ease of fabrication and the fact that useful operation is generally required over a range of operating conditions. The options available regarding doping are well illustrated by the history of IMPATT development. A range of distinguishable doping profile types and the associated field profiles at breakdown are shown in Fig. 16. Profile (a) is the form of the IMPATT as originally proposed by Read [69]. This is designed to follow closely the idealized transit-time concept, with avalanche generation in the narrow high-field region followed by drift at saturated velocities across the drift region. Until quite recently this structure was difficult to fabricate. It was found that quite easily fabricated, uniformly doped structures (b) gave reasonable performance (10–15% efficiency at X band) despite introducing velocity unsaturation and losses in epitaxial material. A variant of the uniformly doped structure giving improved performance is the double-drift structure (c) in which electrons and holes each have a drift region and contribute to the induced current, and the dc voltage is significantly higher than for single-drift structures. Typical efficiencies of 18% at X band are achievable.

Structures (b) and (c) employ doping levels that must simultaneously satisfy the drift region length requirement and the ionization region field requirements. As far as each individually is concerned, it is to be expected that the doping level is not optimum. The attraction of submicron technology is in the opportunity it provides for individually optimizing the

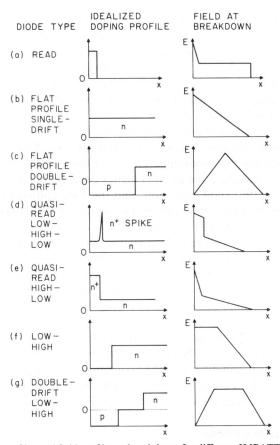

Fig. 16. Doping profiles and field profiles at breakdown for different IMPATT structures.

injection and drift regions. This proved to be very important to IMPATT diodes when quasi-Read structures (d) and (e), fabricated from GaAs, were found to give efficiencies of typically 25–30% and as much as 36% [92] at X band. These structures are termed "quasi"-Read structures because one aspect of "true" Read diodes, carrier flow at high-field saturated velocities, must be avoided to obtain the best results. A variety of sometimes rather subtle reasons are responsible for the high efficiency and other distinctive performance characteristics of quasi-Read GaAs IMPATTs; reviews of these in the drift and diffusion approximation are given elsewhere [93,94]. In addition, the various effects associated with the details of the band structure of GaAs, which were referred to in connection with transferred-electron devices, are present in these devices.

Design principles and performance estimates for quasi-Read structures are often claimed to be based on minimizing the avalanche region to total voltage ratio. It seems not to be widely appreciated that this guideline is based on an idealized picture of the true Read structure, and one of the limits imposed is that the velocity not fall out of saturation [95]. In GaAs quasi-Read structures, the fact that the velocity falls out of saturation is specifically exploited and the design principles consequently become rather different [93,94].

Once the limits of the conventional design guideline are appreciated, the philosophy behind structures (f) and (g), low–high and double-drift low–high diodes, becomes apparent. Here the aim is to maximize power. The wider avalanche zone permits higher currents before avalanche resonance effects [96] spoil operation, and the high epitaxial doping permits high voltage modulation depths before epitaxial losses spoil device performance. The high drift region doping is also useful in allowing higher current densities before space-charge compensation of background doping occurs and in reducing the sensitivity to operating conditions which is introduced by depletion region width modulation. It is believed that this design approach, using Si diodes, offers the potential for achieving particularly high powers (if not efficiencies) throughout the useful frequency range for IMPATT operation.

The principle of separation of the injection and drift region requirements also applies to TUNNETTs. Elta and Haddad [68] showed (in the limit of the drift and diffusion approximation) that operation of uniformly doped TUNNETTs will be band-limited and that for operation at lower frequencies the high–low structure (e) of Fig. 16 is required [68].

C. Material Control

It is also possible to improve the characteristics of the injection and drift regions by changing the material, rather than doping, from region to region. Also, the drift process may have different characteristics as a function of phase in the rf cycle by having different materials present within the drift region. The permissible material combinations are limited by the needs for a good match of lattice constants and thermal expansion coefficients [97]; also, there are generally rather high saturation current densities associated with interface states, and these must not spoil device operation. In this section the various approaches to exploiting heterostructures will be outlined in the simplest possible approximation; it will be assumed that the bulk drift and diffusion properties apply within each material. The general approximations inherent in this approach were explained in Section II.B; effects specific to charge transport across heterostructure interfaces may also be important [97].

One concept that is applicable to all pure transit-time devices is velocity modulation in the drift region. In the waveform of Fig. 14b all induced current before the π phase point is lossy; between π and 2π, the region around $3\pi/2$ is the most important. Output power and efficiency can be significantly improved by reducing the current before the π point as much as possible and shifting the center of gravity of the induced current pulse as close as possible to $3\pi/2$. This is achieved if carriers start drifting in a material with a low saturated velocity and sometime after the π point transfer to a material with a higher saturated velocity (Fig. 17). One attractive material combination for this purpose appears to be $GaAl_xAs_{1-x}$(low saturated velocity)–GaAs(high saturated velocity).

The other technique, to optimize the properties of the injection region and drift region using different materials, has been attempted with IMPATTs. The most frequently suggested combination is Ge(avalanche region)–GaAs(drift region). There are at least two reasons to believe that this could be a useful combination. One is associated with the anomalously long avalanche response time in GaAs, which constitutes a high-frequency, performance-limiting mechanism [68]. Use of Ge with its "normal" response time in the avalanche region could permit higher frequency operation. The second reason has to do with the fact that at a given field the ionization coefficients are much lower in GaAs than in Ge. This means that drift region ionization, an important performance- and stability-limiting mechanism [73,98], becomes significant at larger voltage modulation depths so better performance may be achievable [73].

The theoretical advantages of heterostructure IMPATTs have not yet been demonstrated experimentally. Namordi et al. [99] fabricated some Ge–GaAs structures but obtained relatively poor microwave performance (efficiencies <7% at frequencies in the range of 10 to 16 GHz). The reason for this may be the high saturation currents associated with

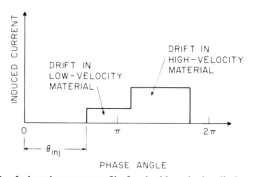

Fig. 17. Induced current profile for double-velocity diode operation.

interface states at the Ge–GaAs junction. Saturation currents are known to be an important IMPATT performance-limiting mechanism [100,101].

If the high saturation current is the reason for the poor performance, there are some important consequences. First, the effects of saturation currents on IMPATT operation become relatively less important as the frequency increases (because modulation of the particle current is less at higher frequencies), so relatively leaky Ge–GaAs IMPATTs could still be useful at millimeter-wave frequencies. Second, TUNNETT operation is almost unaffected by saturation currents, so heterostructure TUNNETTs remain attractive. A three-layer, or "sandwich," double-drift heterostructure TUNNETT is particularly interesting. The central layer would be chosen to optimize the tunneling injection and initial low-velocity drift while the outer layers would be chosen for high-velocity electron and hole transport, respectively.

D. Passive Devices

Whatever the state of the art with respect to the highest frequencies at which active devices can generate power, it is possible to obtain power at higher frequencies by utilizing the nonlinearity of driven passive devices to provide frequency multiplication. Mixing and detection are also achievable using the nonlinearity of passive devices. Various types of diodes can provide the required nonlinearity. Schottky barrier diodes, consisting of a Schottky contact on an epitaxial layer with a heavily doped back substrate, are most commonly used. Schottky barrier diodes with diameters of 1 μm have been used as frequency multipliers in the 450–600-GHz range [102] and as mixers and detectors at submillimeter wavelengths [103,104]. A major problem with the Schottky diode structure is series loss in the undepleted epitaxial material. The Mott diode, a Schottky diode structure with an epitaxial layer designed to be completely depleted, or punched through, at the operating point overcomes this problem. Keen et al. [105] have discussed the use of Mott diode mixers. The main disadvantage of the Mott diode is the tight control required on doping and epitaxial layer width. These limitations can be partially overcome by using a reverse-biased Schottky diode as a replacement for the substrate, forming a BARITT structure [106].

The device diameters associated with submillimeter wavelength diodes are of the order of a micron, showing that submicron technology can be required for defining areas, as well as thicknesses, of devices. The areas of microwave, millimeter-, and submillimeter-wave diodes are limited by a number of considerations, which apply to both active and passive devices. These are thermal considerations, rf impedance matching limitations, and resistive losses. The thermal and rf impedance matching limitations are quite well known. Energy dissipated as heat in a device must be removed without too great a temperature rise or the diode will suffer performance degradation or destruction. Thermal limitations are usually most important for power devices at microwave frequencies. Radio-frequency impedance matching limitations arise because, at microwave and higher frequencies, it is very difficult to match impedances significantly less than approximately 1 Ω. Radio-frequency impedance matching is often the main area-limiting mechanism at millimeter-wave frequencies. At submillimeter-wave frequencies, losses associated with substrate layers and contacts become dominant. In particular, skin effects become important and can greatly increase the effective contact resistance [107]. When considering skin effects in millimeter- and submillimeter-wave diodes, it is important to take into account displacement currents and the intercollisional scattering times of the carriers. Champlin and Eisenstein [108] calculate cutoff frequency versus contact radius for Schottky diodes, taking these effects into account. Similar calculations must also be performed for transit-time devices. Ohmori *et al.* [109] performed a somewhat idealized analysis of the maximum oscillation frequency versus the diameter of IMPATTs.

There are difficult problems associated with mounting and connecting to the small-area diodes required for millimeter- and submillimeter-wave operation. Submicron technology should help to ease this problem by permitting integration of small-area diodes, signal coupling, and bias feeds onto a single piece of semiconductor material. Millimeter-wave integrated circuits should lower the cost, increase the reliability, and potentially improve the performance of subsystems for generation and processing.

E. State of the Art Performance

The current state of the art for microwave, millimeter-wave, and submillimeter-wave power generation [102,110–119] is shown in Fig. 18. It should be remembered that thermal limitations on transit-time diodes give an output power that varies approximately as (frequency)$^{-1}$, and rf impedance limitations give an output power that varies approximately as (frequency)$^{-2}$. When series resistance and skin-effect losses dominate, the power–frequency relationship will fall off even more rapidly. Pulsed results will be better than cw results for the range of conditions where cw operation is thermally limited and pulsed operation is impedance-limited. Over the frequency range considered in Fig. 18, IMPATTs are virtually

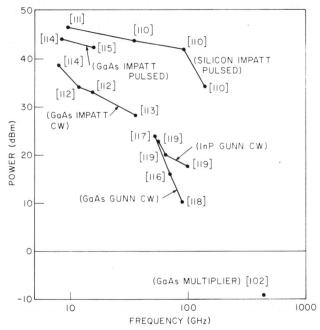

Fig. 18. State of the art for microwave, millimeter-wave and submillimeter-wave power generation using solid-state sources.

without competition as power sources. For cw operation up to 40 GHz, GaAs IMPATTs give the highest output power; for pulsed operation, Si double-drift IMPATTs give the highest output power. Gunn diodes are not competitive as far as output power is concerned, but their lower noise may be useful for some applications up to 100 GHz, with InP devices possibly being useful at somewhat higher frequencies. The best prospects for generating useful amounts of power between 200 and 500 GHz seem to lie with TUNNETT structures.

V. MICROWAVE DEVICES MADE POSSIBLE BY SUBMICRON TECHNOLOGY

The previous two sections of this chapter were concerned with the impact of submicron technology on existing devices, i.e., the effects of scaling down the dimensions of known devices. In this section devices that have only become possible as a result of submicron technology will be considered. The ability to realize such devices is due to very recent improvements in heterostructure fabrication (and, to some extent, theory), and the application to devices is very much in its infancy. The devices proposed to date involve superlattices. In this section a review of basic superlattice principles will be followed by a discussion of the potential application in devices.

A. Review of Superlattice Principles

A superlattice is a structure consisting of thin layers of semiconductor material with the material type, thickness, and doping individually controlled from layer to layer. The layer thickness can be as small as 100 to 200 Å, a dimension small enough to impose quantum effects on carrier transport. Most superlattices are fabricated using molecular-beam epitaxy (MBE). This technique [1] is particularly well suited to superlattice fabrication for several reasons. It is a low-temperature ($\approx 600°C$) epitaxial growth process in which individual beams of various elements are directed onto a crystalline surface under vacuum conditions. Since the individual beams can be rapidly modulated by means of shutters, the material type and stoichiometry of the growth can be rapidly changed. The growth rate can be controlled by varying the temperature of the beam sources, and doping can be controlled with a separate impurity beam. Layers can be selectively doped using a shutter in this impurity beam. Because of the low-growth temperature there is little diffusion between layers. However, MBE is not a prerequisite for superlattice fabrication; it is also possible to use liquid phase epitaxial growth [120].

Many features of carrier transport in these thin semiconductor layers may be understood by assuming that normal band theory applies within each layer. The energy band discontinuities may be described in terms of idealized heterojunction theory [97]. Material types are chosen to give the desired band properties and are, of course, limited by the practical considerations (matching lattice constants and thermal expansion coefficients) noted in Section IV. Among the practical material combinations are Ge–GaAs [99], GaAs–$Ga_xAl_{1-x}As$ [121,122], and InP–$GaP_{1-z}As_z$ [123]. Various effects associated with superlattices may be predicted on the basis of the idealized superlattice band edge versus distance profile of Fig. 19.

First, it is to be expected that the quantization of energies associated with directions perpendicular to the layers will become significant, although energies associated with motion parallel to the layers will remain effectively continuous. (Energy levels are effectively continuous if their separation is much less than the mean thermal energy of the carriers.) For

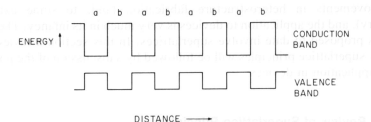

ENERGY

Fig. 19.   Ideal band-edge diagram for a superlattice.

a deep one-dimensional potential well the bound states can be approximated by those of an infinite potential well, which are [124]

$$E_n = (\pi\hbar^2/2m^*a^2)n^2,$$

where E_n is the energy of the nth state above (below) the conduction (valence) band edge, m^* the electron (hole) effective mass, $\hbar$ Planck's constant divided by 2π, and a the width of the potential well. To provide an idea of the dimensions for which quantum effects become important, it may be noted that if m^* is taken as the free electron mass, the separation between the first and second energy levels is 10 meV when $a \simeq 150$ Å. In addition to the size quantization of energy levels, the narrow spatial dimensions also introduce a stepped density of states for the perpendicular direction. Although such quantization will affect the transport properties, as in narrow inversion layers in MOS devices [30–32], it does not appear to have been suggested that exploitation of the quantization could form the basis of a useful device.

A second effect is that electrons present in the superlattice due to donor impurities in material b of Fig. 19 will tend to fall into the lower energy states associated with the layers of material a. This gives the possibility of obtaining high mobilities in material a by avoiding impurity scattering, since the donor impurities are in material b. This is termed "modulation doping."

A third effect is that although in nondegenerate material electrons will predominantly be in the low energy states in material a at thermal equilibrium, when they are made more energetic (e.g., by an applied electric field), they will have a tendency to transfer to the higher energy states associated with material b. If the mobility in material b is lower than the mobility in material a, a negative differential mobility may arise. The process affecting the mobility is motion in real space and is termed "real-space transfer" (RST). This is in contrast to the negative differential mobility associated with the Gunn effect (Section IV), which is due to transfer of electrons in k space.

With this background it becomes possible to explain at least the basic principles of proposed devices based on superlattices. This will now be done, first for devices exploiting transport predominantly parallel to the superlattice layers, and then for devices exploiting transport perpendicular to the layers.

B. Parallel Transport and Real-Space-Transfer Devices

The negative differential mobility associated with real-space transfer should be exploited analogously to the ways in which the negative differential mobility is exploited in bulk GaAs and InP. It has been proposed that this would result in a useful microwave device with a cutoff frequency estimated as 200 GHz [125,126]. Although the basic principle of the negative-resistance RST device seems valid, it is difficult to argue that the device would have properties in any way superior to traditional transferred-electron devices. Performance would probably be inferior since the energy and momentum relaxation effects which limit traditional transferred-electron devices are still present, and in addition there are the diffusion-related limitations associated with the real-space-transfer process. On this basis it seems likely that 40 GHz may represent a more realistic upper frequency limit and that performance characteristics (output power and efficiency in particular) will be inferior to conventional transferred-electron devices, even at lower frequencies. Some thought has been given to using a third contact in conjunction with the basic real-space-transfer structure. The third contact could be arranged in such a way as to permit an input signal at the third terminal to establish an rf field perpendicular to the basic (parallel) transport direction, thereby enhancing the basic real-space-transfer mechanism. This principle, termed field-enhanced real-space transfer (FERST), is currently being analyzed by the authors.

Modulation doping has been demonstrated experimentally [127]. However, the reduction in impurity scattering only leads to a useful increase in mobility in circumstances where impurity scattering would be an important scattering mechanism, i.e., at high doping levels or at low temperatures, and for low-energy electrons [4]. At low fields the basic system yields what is essentially a low-loss ohmic conductor, which is not particularly useful. At higher fields, as electrons become on the average more energetic, the real-space-transfer effect would become significant, removing the high mobility properties. (A modulation doped structure would probably be a good choice for a real-space-transfer device if the

RST approach were pursued.) It may be that an ingenious structure could combine the concept of high mobility from modulation doping with the ballistic transistor principles discussed in Section III, but the ballistic transistor principle itself has yet to be verified by experiment.

C. Perpendicular (Sublattice) Transport

The idea of exploiting perpendicular (sublattice) transport seems to predate the ideas of exploiting perpendicular transport and real-space transfer, but the theoretical ideas have yet to be demonstrated experimentally. The basic idea is to exploit the band structure of the superlattice to achieve a situation where many of the carriers are at a negative effective mass part of the $E-k$ characteristic for motion perpendicular to the superlattice [128,129]. If sufficient carriers have negative effective masses, a negative mobility may result, which could be used to produce a device with negative-resistance characteristics.

The reason that this situation can, in principle, be achieved in a superlattice structure is that for superlattices the Brillouin zone is split into subzones by the double periodicity of the lattice and the superlattice. These subzones in turn result in subbands in the $E-k$ diagram. A set of subbands is shown in Fig. 20. Since the slopes of the $E-k$ curves are zero at the minizone boundaries, a region of negative effective mass occurs within each

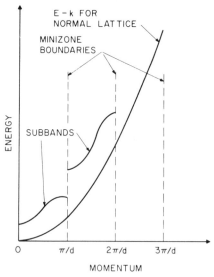

Fig. 20.   Energy–momentum diagram for perpendicular superlattice transport.

subzone. Thus at quite low energies, carriers may have negative effective masses. (In a bulk semiconductor, in contrast, the energies associated with negative effective masses are much higher and it does not seem possible to produce a situation in which the majority of carriers have negative effective masses.)

Clearly the overall transport properties are determined by the combination of scattering processes and applied electric field, as well as the $E-k$ characteristic. Theory predicts useful device operation for a variety of realistic assumptions regarding scattering processes [128,129], but experimental confirmation has not been achieved. This could be due to a number of factors, among them "smearing" of the $E-k$ relationship due to fabrication tolerances reducing the assumed periodicity and the space-charge instabilities to be expected in a negative mobility material.

D. Discussion

Overall, there does not yet appear to be any compelling reason to believe that superlattice devices will make a significant impact in the foreseeable future. The real-space transfer and modulation doping concepts both seem to work but not to offer major performance improvements compared to more traditional approaches. Perpendicular transport has not yet been demonstrated experimentally, and even if it were, device performance might be expected to be very temperature sensitive through the temperature dependence of the scattering rates.

Despite this pessimistic assessment, this section serves a valuable purpose insofar as it shows that completely new device principles are possible. There is often a lag between "fundamental" research and practical applications, and it would certainly be shortsighted to assume that there will be no substantial application of superlattice material technology. Also, as noted in Section II, it will, in the future, be necessary for device theorists to look beyond "traditional" device concepts. Consideration of the exploitation of superlattices is a step in that direction.

VI. DISCUSSION AND CONCLUSIONS

In this final section the consequences of submicron technology for microwave and millimeter-wave devices will be divided into two categories, those which are already apparent and those which are more speculative but are predicted to take place in the future.

The most significant impact to date is the capability for making high-performance, submicron geometry, field-effect transistors, especially GaAs MESFETs. Availability of these devices, whose principles of operation and properties were discussed in Section III, is greatly altering the pattern of use of active semiconductor devices at frequencies below approximately 30 GHz. Where previously a range of devices (IMPATTs, TEDs, BARITTs) were used depending on noise and power requirements, submicron geometry FETs are, or soon will be, natural choices for most applications requiring anything other than the highest available output powers. Not only do these devices have excellent low-noise characteristics and good output power and efficiency capabilities, but they also have the circuit advantages associated with separation of input and output functions in three (or more) terminal devices. Also, as noted in Section III, FETs have various other advantages such as self-ballasting and resistance to radiation effects.

There are few obvious niches for other semiconductor devices at frequencies up to 30 GHz. Some, having simpler structures, are less expensive to fabricate than FETs and so offer cost advantages. Gunn and BARITT devices, for instance, will continue to be useful for high-volume, low-cost items such as intrusion alarms and doppler systems. However, in many applications device cost is only a small fraction of overall system cost, and in any case the cost of microwave FETs should fall significantly in the future. For high-power applications where noise requirements are not too stringent, some of the multilayer IMPATTs discussed in Section IV may be most suitable.

At frequencies in the range of 30 to 200 GHz, Si double-drift IMPATTs currently offer the best output power and efficiency performance. In the lower half of the frequency range, transferred-electron devices may be suitable for some relatively low-noise, low-power applications. It remains to be seen whether, with improvements in device design and technology, FETs will be able to dominate the lower part of this frequency range as they almost certainly will up to 30 GHz.

At higher frequencies still, IMPATT performance drops off significantly, as discussed in Section IV, and the TUNNETT should become the most attractive transit-time device. It should combine low noise with reasonable output power and efficiency. If, or more optimistically when, it becomes possible to fabricate single- and double-drift heterostructure TUNNETTs, surprisingly high output powers and efficiencies may be achieved. Such devices should prove extremely useful throughout the millimeter- and submillimeter-wave regions.

Submicron technology will impact increasingly on subsystem integration. Integrating the functions of power generation, amplification, mixing, detection, control, and (when appropriate) coupling to optical components offers many potential advantages including cost, size, reliability, and performance. Integration at X-band frequencies is already being achieved and submicron technology should permit extension of the techniques to millimeter- and even submillimeter-wave frequencies.

It is not clear that any of the currently proposed superlattice and real-space-transfer devices made possible by submicron technology are capable of performances competitive with the more traditional devices whose dimensions are merely made smaller using submicron technology. Section V is included for completeness and in the hope of provoking further thought on the exploitation of existing technological capabilities.

In conclusion, the obvious impact to date of submicron technology on microwave and millimeter-wave devices has been in permitting greatly improved performance from field-effect transistors, especially GaAs MESFETs, at frequencies where previously a variety of two-terminal negative-resistance devices were used. During the next few years the impact may be in two areas: improved millimeter- and submillimeter-wave power generation capabilities and integration of individual components into single-chip subsystems. Various TUNNETT structures appear highly promising in the first of these areas, but it is possible that conceptually completely new devices will be discovered. Much experimental and theoretical work remains to be done.

REFERENCES

1. M. G. Panish and A. Y. Cho, *IEEE Spectrum* **17**, 18 (1980).
2. G. Carter and W. A. Grant, "Ion Implantation of Semiconductors." Wiley, New York, 1976.
3. R. C. Henderson, D. C. Mayer, and J. G. Nash, *J. Vac. Sci. Technol.* **16**, 260 (1979).
4. E. M. Conwell, High field transport in semiconductors, *Solid State Phys. Suppl.* **9** (1967).
5. P. J. Price, *Solid-State Electron.* **21**, 9 (1978).
6. D. K. Ferry, J. R. Barker, and C. Jacoboni (eds.), "Physics of Nonlinear Transport in Semiconductors," Nato Advanced Study Institutes Series B: Physics, Vol. 52. Plenum Press, New York, 1980.
7. J. R. Barker, *in* "Physics of Nonlinear Transport in Semiconductors" (D. K. Ferry, J. R. Barker, and C. Jacoboni, eds.), Nato Advanced Study Institutes Series B: Physics, Vol. 52, Lecture 5. Plenum Press, New York, 1980.
8. W. Fawcett, A. D. Boardman, and S. Swain, *J. Phys. Chem. Solids* **31**, 1963 (1970).
9. P. J. Price, *in* "Semiconductors and Semimetals" (R. K. Willardson and A. C. Beer, eds.), Vol. 14, Chapter 4. Academic Press, New York, 1979.
10. H. D. Rees, *J. Phys. Chem. Solids* **30**, 643 (1969).
11. P. A. Lebwohl and P. J. Price, *Appl. Phys. Lett.* **19**, 530 (1971).
12. M. Abe, S. Yanagisawa, O. Wada, and H. Takanishi, *Jpn. J. Appl. Phys.* **14**, 70 (1975).

13. R. A. Warriner, *IEE J. Solid State Electron Devices* **1**, 97 (1977).
14. R. A. Warriner, *Solid State Electron Devices* **1**, 105 (1977).
15. K. Blotekjaer, *IEEE Trans. Electron Devices* **ED-17**, 38 (1970).
16. W. Fawcett, *in* "Electrons in Crystalline Solids." IAEA, Vienna, 1973.
17. P. N. Butcher, *Rep. Prog. Phys.* **30**, 97 (1967).
18. K. Blotekjaer and E. B. Lunde, *Phys. Status Solidi* **35**, 581 (1969).
19. P. N. Butcher and C. J. Hearn, *Electron. Lett.* **4**, 459 (1968).
20. R. Bosch and H. W. Thim, *IEEE Trans. Electron Devices* **ED-21**, 16 (1974).
21. R. Froelich, The Univ. of Michigan, private communication (1980).
22. A. F. Gibson, J. W. Granville, and E. G. S. Paige, *J. Phys. Chem. Solids* **19**, 198 (1961).
23. M. S. Shur, *Electron. Lett.* **12**, 615 (1976).
24. A. Kaszynski, Étude des Phenomenes de Transport dans les Materiaux Semiconducteurs par les Methodes de Monte Carlo: Application a L'Arseniure de Gallium de Type *N*. Ph.D. Thesis, Univ. of Lille (1979).
25. I. Doumbia, G. Salmer, and E. Constant, *J. Appl. Phys.* **46**, 1831 (1975).
26. B. Carnez, A. Cappy, A. Kaszynski, E. Constant, and G. Salmer, *J. Appl. Phys.* **51**, 784 (1980).
27. J. Pribetich, M. Lefebvre, and E. Allamando, *Electron. Lett.* **12**, 460 (1976).
28. P. A. Blakey and J. R. East, Improved modeling of GaAs IMPATTs, *Proc. Biennial Conf. Active Microwave Semicond. Devices Circuits, 7th, Ithaca, New York* (1979).
29. C. B. Duke, "Tunneling in Solids." Academic Press, New York, 1969.
30. F. Stern, *CRC Crit. Rev. Solid State Sci.* **4**, 499 (1974).
31. D. K. Ferry, *Solid-State Electron.* **21**, 115 (1978).
32. P. J. Stiles, Transport in quasi-two-dimensional space charge layers (invited paper), *Proc. Int. Conf. Phys. Semicond. 14th* Conf. Ser. Number 43, p. 41. Institute of Physics, London, 1979.
33. R. Allison, *IEEE Trans. Microwave Theory Tech.* **MTT-27**, 415 (1979).
34. J. G. Oakes, R. A. Wickstrom, D. A. Tremere, and T. M. S. Heng, *IEEE Trans. Microwave Theory Tech.* **MTT-24**, 305 (1976).
35. C. Tsironis and U. Niggebrügge, *IEEE Trans. Microwave Theory Tech.* **MTT-27**, 1052 (1979).
36. W. Shockley, *Proc. IRE* **40**, 1365 (1952).
37. R. Pucel, H. Haus, and H. Statz, *Adv. Electron. Electron Phys.* **38**, 195 (1975).
38. C. A. Liechti, *IEEE Trans. Microwave Theory Techn.* **MTT-24**, 279 (1976).
39. J. V. DiLorenzo and W. R. Wisseman, *IEEE Trans. Microwave Theory Tech.* **MTT-27**, 367 (1979).
40. D. P. Kennedy and R. R. O'Brien, *IBM J. Res. Dev.* **14**, 95 (1970).
41. B. Himsworth, *Solid-State Electron.* **15**, 1353 (1972).
42. B. Himsworth, *Solid-State Electron.* **16**, 931 (1973).
43. M. Reiser, *IEEE Trans. Electron Devices* **ED-20**, 35 (1973).
44. J. J. Barnes and R. J. Lomax, *IEEE Trans. Electron Devices* **ED-24**, 1082 (1977).
45. T. Adachi, A. Yoshii, and T. Sudo, *IEEE Trans. Electron Devices* **ED-26**, 1026 (1979).
46. J. R. East and N. A. Masnari, An efficient pseudo-two-dimensional simulation of MESFET operation, *Proc. Biennial Cornell Elec. Eng. Conf., 7th, Ithaca, New York* (1979).
47. J. G. Ruch, *IEEE Trans. Electron Devices* **ED-19**, 652 (1972).
48. T. J. Maloney and J. Frey, *J. Appl. Phys.* **48**, 781 (1977).
49. S. Kratzer and J. Frey, *J. Appl. Phys.* **49**, 4064 (1978).
50. C. O. Bozler, G. O. Alley, R. A. Murphy, and D. E. Flanders, Permeable base transistor, *Proc. Biennial Cornell Elec. Eng. Conf., 7th, Ithaca, New York* (1979).

51. G. O. Alley, C. O. Bozler, R. A. Murphy, and M. T. Lindley, Two-dimensional numerical simulation of the permeable base transistor, *Proc. Biennial Cornell Elec. Eng. Conf. 7th, Ithaca, New York* (1979).
52. M. S. Shur and L. F. Eastman, *IEEE Trans. Electron Devices* **ED-26,** 1677 (1979).
53. T. J. Maloney, *IEEE Electron Device Lett.* **EDL-1,** 54 (1980).
54. J. M. Schellenburg, G. Keithley, C. K. Kim, V. K. Eu, H. B. Kim, and G. O. Ladd, RF characteristics of Ka-Band GaAs FETs, *Proc. Biennial Cornell Elec. Eng. Conf., 7th, Ithaca, New York* (1979).
55. C. F. Krumm, H. T. Suyematsu, and B. L. Walsh, A 30 GHz GaAs FET amplifier, *Int. Microwave Symp. Digest, Ottawa, Canada* p. 385 (1978).
56. T. Suzuki, A. Nara, M. Nakatani, and T. Ishii, *IEEE Trans. Microwave Theory Tech.* **MTT-27,** 1070 (1979).
57. G. O. Ladd *et al.,* Ku band, low noise GaAs FETs, *Proc. Biennial Cornell Conf. Elec. Eng., 6th, Ithaca, New York* p. 369 (1977).
58. W. Shockley, *Bell Syst. Tech. J.* **33,** 799 (1954).
59. S. Ramo, *Proc. IRE* **27,** 584 (1939).
60. W. Shockley, *Bell Syst. Tech. J.* **30,** 990 (1951).
61. D. J. Coleman and S. M. Sze, *Bell Syst. Tech. J.* **50,** 1695 (1971).
62. G. T. Wright, *Solid-State Electron.* **16,** 903 (1973).
63. J. R. East, H. Nguyen-Ba, and G. I. Haddad, *IEEE Trans. Microwave Theory Tech.* **MTT-24,** 943 (1976).
64. S. P. Kwok and G. I. Haddad, *Solid-State Electron.* **19,** 795 (1976).
65. J. Freyer and P. N. Fong, *Proc. IEE* **127,** Part I, 78 (1980).
66. J. Nishizawa, T. Ohmi, and T. Sakai, Millimeter-wave oscillations from TUNNETT diodes, *Proc. Eur. Microwave Conf., 4th, Montreux, Switzerland* p. 449 (1974).
67. J. Nishizawa, K. Motoya, and Y. Okuno, *Jpn. J. Appl. Phys. Suppl. 17-1* **17,** 167 (1978).
68. M. E. Elta and G. I. Haddad, *IEEE Trans. Microwave Theory Tech.* **MTT-27,** 442 (1979).
69. W. T. Read, *Bell Syst. Tech. J.* **37,** 401 (1958).
70. G. I. Haddad, P. T. Greiling, and W. E. Schroeder, *IEEE Trans. Microwave Theory Tech.* **MTT-18,** 752 (1970).
71. G. I. Haddad (ed.), "Avalanche Transit-Time Devices." Artech House, Dedham, Massachusetts, 1972.
72. G. Gibbons, "Avalanche-Diode Microwave Oscillators." Oxford Univ. Press, London and New York, 1973.
73. B. Culshaw, R. A. Giblin, and P. A. Blakey, "Avalanche Diode Oscillators." Taylor and Francis, London, 1978.
74. M. E. Elta and G. I. Haddad, *IEEE Trans. Electron Devices* **ED-25,** 694 (1978).
75. M. E. Elta and G. I. Haddad, *IEEE Trans. Electron Devices* **ED-26,** 941 (1979).
76. S. Ahmad and W. S. Khokley, *J. Phys. Chem. Solids* **28,** 2499 (1967).
77. C. Shekhar and S. K. Sharma, *Phys. Lett.* **50A,** 120 (1974).
78. C. Shekhar and S. K. Sharma, *J. Phys. C: Solid State Phys.* **8,** 4249 (1975).
79. M. Takeshima, *Phys. Rev.* **B 12,** 575 (1975).
80. D. Hill, *J. Phys. C: Solid State Phys.* **9,** 3527 (1976).
81. J, J. Berenz, J. Kinoshita, T. L. Hierl, and C. A. Lee, *Electron Lett.* **15,** 150 (1979).
82. T. P. Pearsall, F. Capasso, R. E. Nahory, M. A. Pollak, and J. R. Chelikowsky, *Solid-State Electron.* **21,** 297 (1978).
83. H. D. Law and C. A. Lee, *Solid-State Electron.* **21,** 331 (1978).
84. H. Shichijo, K. Hess, and G. E. Stillman, *Electron. Lett.* **16,** 208 (1980).
85. R. Ghosh and S. K. Roy, *Solid-State Electron.* **18,** 945 (1975).

86. B. B. van Iperen and H. Tjassens, *Proc. IEEE* **59**, 1032 (1971).
87. L. F. Eastman (ed.), "Gallium Arsenide Microwave Bulk and Transit-Time Devices." Artech House, Dedham, Massachusetts, 1972.
88. B. G. Bosch and R. W. H. Engelman, "Gunn-Effect Electronics." Wiley, New York, 1975.
89. P. J. Bulman, G. S. Hobson, and B. C. Taylor, "Transferred Electron Devices." Academic Press, New York, 1972.
90. D. Jones and H. D. Rees, *J. Phys. C: Solid-State Phys.* **6**, 1781 (1973).
91. H. D. Rees and K. W. Gray, *IEE J. Solid-State Electron Devices* **1**, 1 (1976).
92. C. K. Kim, W. G. Matthei, and R. Steele, "GaAs Read IMPATT diode oscillators, *Proc. Biennial Cornell Elec. Eng. Conf., 4th, Ithaca, New York* p. 209 (1974).
93. P. A. Blakey, B. Culshaw, and R. A. Giblin, *IEEE Trans. Electron Devices* **ED-25**, 674 (1978).
94. P. E. Bauhahn and G. I. Haddad, *IEEE Trans. Electron Devices* **ED-24**, 634 (1977).
95. D. L. Scharfetter and H. K. Gummel, *IEEE Trans. Electron Devices* **ED-16**, 64 (1969).
96. M. Gilden and M. E. Hines, *IEEE Trans. Electron Devices* **ED-13**, 169 (1966).
97. A. G. Milnes and D. L. Feucht, "Heterojunctions and Metal–Semiconductor Junctions." Academic Press, New York, 1972.
98. B. B. van Iperen, *Proc. IEEE* **62**, 284 (1974).
99. M. R. Namordi, D. W. Shaw, and F. H. Doerbeck, *IEEE Trans. Electron Devices* **ED-26**, 1074 (1979).
100. T. Misawa, *Solid-State Electron.* **13**, 1363 (1970).
101. D. R. Decker, C. N. Dunn, and H. B. Frost, *IEEE Trans. Electron Devices* **ED-18**, 141 (1971).
102. T. Takada and M. Ohmori, *IEEE Trans. Microwave Theory Tech.* **MIT-27**, 519 (1979).
103. *IEEE Trans. Microwave Theory Tech.* **MTT-25**, No. 6 (1977), Special Issue on the *Proc. Int. Conf. Submillimeter Waves Their Appl. 2nd.*
104. W. M. Kelly and G. T. Wrixon, *IEEE Trans. Microwave Theory Tech.* **MTT-27**, 665 (1979).
105. N. J. Keen, R. W. Haas, and E. Perchtold, *Electron. Lett.* **14**, 825 (1978).
106. P. J. McCleer and G. I. Haddad, BARITT diode video detectors, *MTT Symp. Digest, Ottawa, Canada* p. 372 (1978).
107. L. E. Dickens, *IEEE Trans. Microwave Theory Techniques* **MTT-15**, 101 (1967).
108. K. S. Champlin and G. Eisenstein, *IEEE Trans. Microwave Theory Tech.* **MTT-26**, 31 (1978).
109. M. Ohmori, T. Ishibashi, and S. Ono, *IEEE Trans. Electron Devices* **ED-24**, 1323 (1977).
110. T. T. Fong and H. J. Kuno, *IEEE Trans. Microwave Theory Tech.* **MTT-27**, 492 (1979).
111. G. Pfund, *IEEE Trans. Microwave Theory Tech.* **MTT-27**, 450 (1979).
112. P. Brook, J. G. Smith, L. D. Clough, C. A. Tearle, G. Ball, and J. C. H. Birbeck, Design, fabrication and performance of GaAs high-efficiency IMPATT diodes in J-band (Ka-band), *Proc. Biennial Cornell Elec. Eng. Conf., 6th, Ithaca, New York* p. 221 (1977).
113. M. G. Adlerstein, R. N. Wallace, and S. Steele, *IEEE Trans. Electron Devices* **ED-25**, 1151 (1978).
114. R. N. Wallace, S. R. Steele, and M. G. Adlerstein, Performance of GaAs double-drift avalanche diodes, *Proc. Biennial Cornell Conf. Elec. Eng., 6th, Ithaca, New York* p. 195 (1977).
115. T. L. Hierl, J. J. Berenz, and S. I. Long, GaAs pulsed read IMPATT diodes, *Proc. Biennial Cornell Elec. Eng. Conf., 6th, Ithaca, New York* p. 211 (1977).

116. A. K. Talwar, *IEEE Trans. Microwave Theory Techniques* **MTT-27**, 510 (1979).
117. C. Sun, E. Benko, and J. W. Tully, *IEEE Trans. Microwave Theory Tech.* **MTT-27**, 512 (1979).
118. J. Ondria, Wide-band mechanically tunable W-band (75–110 GHz) cw GaAs Gunn diode oscillator, *Proc. Biennial Cornell Conf. Elec. Eng., 7th, Ithaca, New York* (1979).
119. J. D. Crowley, F. B. Fank, S. B. Hyder, J. J. Sowers, and D. Tringali, Millimeter wave InP transferred electron devices, *Proc. Biennial Cornell Conf. Elec. Eng., 7th, Ithaca, New York* (1979).
120. E. A. Rezek, R. Chin, N. Holonyak, Jr., S. W. Kirchoefer, and R. M. Kolbas, *J. Electron. Mat.* **9**, 1 (1980).
121. J. P. van der Ziel, R. Dingle, R. C. Miller, W. Wiegmann, and W. A. Nordland, Jr., *Appl. Phys. Lett.* **26**, 463 (1975).
122. N. Holonyak, Jr., R. M. Kolbas, E. A. Rezek, and R. Chin, *J. Appl. Phys.* **49**, 5392 (1978).
123. E. A. Rezek, B. A. Vojak, and N. Holonyak, Jr., *J. Appl. Phys.* **49**, 5398 (1978).
124. G. Baym, "Lectures on Quantum Mechanics," Chapter 4. Benjamin, New York, 1969.
125. K. Hess, H. Morkoc, H. Shichijo, and B. G. Streetman, *Appl. Phys. Lett.* **35**, 469 (1979).
126. K. Hess, *Appl. Phys. Lett.* **35**, 484 (1979).
127. R. Dingle, H. L. Stormer, A. C. Gossard, and W. Wiegmann, *Appl. Phys. Lett.* **33**, 665 (1978).
128. L. Esaki and R. Tsu, *IBM J. Res. Dev.* **14**, 61 (1970).
129. P. Lebwohl and R. Tsu, *J. Appl. Phys.* **41**, 2664 (1970).

Chapter 5

Microfabrication and Basic Research

W. F. BRINKMAN

Bell Laboratories
Murray Hill, New Jersey

I. INTRODUCTION

The development of integrated circuit technology has its origins in the growth and understanding of solid-state physics in the forties and fifties. Since the beginning, however, it has been dependent on continuous developments in a broad spectrum of science and technology. The importance of research and development to integrated circuit technology is widely recognized; in fact, the industry is often claimed to be the most research-intensive industry today. Indeed, company dominance has frequently been determined by the introduction of innovative circuits that give a slight competitive advantage. As the industry has grown, a broader

149

spectrum of companies throughout the world, but particularly in Japan, has been in competition, so that today integrated circuit technology is perhaps even more competitive.

In such an environment it is often easy to overlook the long-range research developments that have been essential to the origins of an industry. This chapter attempts to discuss the role of long-range research in the microfabrication industry. Long-range research necessarily has a twenty-year, or possibly even longer, time scale, so that the many immediate needs (next five years) of the industry are left to other chapters in this book. Here, in order to give perspective on how basic research on a broad front influences the specific development of various parts of integrated circuit technology today, the histories of two specific examples, ion implantation and MOS inversion layers, will be discussed. The important point is that scientific information forms a broad base from which technology is drawn but also that technological developments feed back and influence research in that they frequently define what is possible. Any specific development is imbedded in and surrounded by science and technology progressing together in a complex fashion, in a community, intermixed with people of varying views and goals.

After discussion of these two historical developments, some examples of how basic research and microfabrication are interwoven today and will continue to be in the near future will be given. A number of specific examples of the research needs of microfabrication will be given, particularly in materials science, and a number of uses of microfabrication in research will be cited. Then, recent advances in research on surfaces and interfaces will be given as examples of research relevant to new technology for the future. In the concluding section, the question of the impact of our longest-range research will be raised and a few speculative proposals will be made.

There are scientists and engineers who may consider the theme of this chapter "old hat," or a rehash of the age-old justification of basic research. Old principles, however, need continuous refurbishment and restatement. Concepts as simple as individual freedom are also "age old" and considerably less subtle than the reasons behind support of basic research. Yet there are large portions of the world where individual freedom is ignored. Indeed, in the world today, technology and science are continually being questioned, and in the late sixties and early seventies even the scientific community strongly questioned its own usefulness. This chapter is, therefore, not presented to a more "popular" front, but rather in a series devoted to microfabrication technology. Need we be reminded that in the late sixties many electronics practitioners questioned why any more semiconductor development should be done?

II. HISTORICAL PERSPECTIVE

Whenever the history of microfabrication of semiconductors is discussed, invariably the intense research period immediately following World War II and the invention of the transistor are cited as the genesis of the field [1]. There is no question that this is the case. The development of the quantum theory of solids, the concepts of energy bands and electrons and holes, the understanding of electronic mobilities and doping, the invention of the $p-n$ junction, and finally the transistor itself do indeed form the basis for our present-day semiconductor industry. However, it took thirty additional years of research and development to arrive at present-day microfabrications capabilities and basic research has contributed to this development in many ways. In this section, the development of two integral parts of microfabrication will be traced historically. These were chosen partly because of the author's familiarity but also because they represented good examples of two quite different ways in which research has contributed to the field. The first, the inversion layer, has grown out of and is part of the mainstream of semiconductor research and development. Its origins start with problems with the transistor itself [2]. The second, ion–solid interactions and ion implantation, has its beginnings in studies of radiation effects and nuclear physics. It represents a set of fabrication and analytical techniques dependent on the understanding of the ion–solid interactions that were first considered with respect to nuclear reactors and particle accelerators. Ion implantation is an excellent example of how cross-disciplinary research can have a major technological impact.

There are many other examples that could be discussed—the development of organic resists, high-purity single-crystal silicon, and electron-beam lithography to name a few. The intention here is only to illustrate how microfabrication technology has developed through its coupling to basic research.

A. Inversion Layers—MOSFET

In the past few years the sales of integrated circuits based on metal–oxide–semiconductor field-effect transistors (MOSFET) have surpassed the billion-dollar level. These devices are all based on the ability to apply a voltage to a metal "gate" (see Fig. 1a), electrically insulated from the silicon semiconductor by SiO_2, and for the voltage to cause a bending of the semiconductor bands until the conduction band edge at the surface is

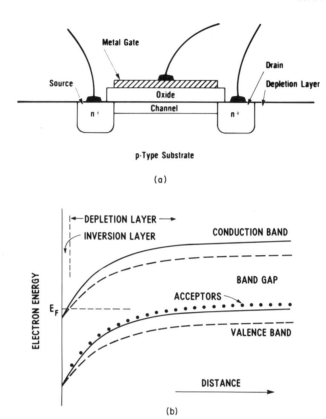

Fig. 1 (a) Configuration of typical metal–oxide–semiconductor field-effect transistor; (b) band bending in inversion layer region of MOSFET.

well below the valence band edge in the bulk (Fig. 1b), assuming the bulk is p-type). The conduction band then will be populated by electrons and the conduction type is said to be inverted. The large modulation of the conductivity at the inversion layer induced by the gate voltage makes the MOSFET an almost ideal amplifier [3]. At the same time, it has turned out that, because the inversion layer is thin enough to be effectively two dimensional and the number of electrons in the inversion layer can be varied widely, this system has become an important system for basic research studies of a two-dimensional electron gas [4]. Thus the MOS device has become an important means for doing research as well as a major technological innovation. Its history and present-day status represent a classic example of the continuous interplay between research and electronics technology, which I shall attempt to relate here.

The MOSFET is, of course, based on the quantum theory of energy bands in solids so that it has its base in the research of the thirties and forties. As seen from Fig. 1a, it is dependent on the inversion of the $p-n$ junction, but it is also dependent on the idea of field-effect devices, which was proposed in the early thirties as well. Lilienfeld in the U.S. [5] and Heil in Britain [6] both had the idea of modulating the conductivity of one material near its surface by applying a voltage on it through an insulating region.

The major thrust in the development of semiconductor electronics, however, came with the work just after World War II, primarily at Bell Laboratories. The field effect [7] in semiconductors was first demonstrated by showing that the majority carrier conductivity could be modulated by an electric field; this led to much activity on such devices.

The invention of the bipolar transistor by Shockley, Bardeen, and Brattain initiated an enormous research and development effort aimed toward its utilization. In this effort a problem with transistors led to some of the first research important to the MOS field-effect transistor. In fabricating transistors it was found that the conductance between the emitter and collector was often much higher than expected. Studies of this effect led to the postulate of a conducting channel on the surface between the two $p-n$ junctions thereby bridging the emitter and collector. The channel was thought to arise because of ions on the surface that repelled majority (p) carriers in the base and attracted minority (n) carriers in an $n-p-n$ transistor structure. Because the emitter and collector are n-type, the n-type surface layer would act as a conducting short between them. The self-consistent band bending that is implied by this model was first analyzed by Brown [8], who showed it to be experimentally correct. Today the region of band bending where a channel for minority carriers is created is called an inversion layer. The inversion layer itself was born from research to better understand the workings of the transistor. Shortly after this the idea that the inversion layer could be induced electrically by an electrode near the base of a transistor and used as an FET was patented by Ross [9]. In this patent the insulating layer was to be a ferroelectric that would be switched by applying voltages to a gate—a rather intimate field effect, but one that had the advantage of memory.

It was recognized at the time that an insulating layer with a metallic "gate" would work instead of a ferroelectric. Such devices, however, were not realizable because of the lack of an insulating layer thin enough to allow the field effect but of sufficient dielectric strength to withstand the fields of 10^6 V/cm necessary to obtain high conductivity channels. In addition, the interface between the silicon and the insulating layer had to have a small enough number of interfacial electronic states so that their filling and emptying did not pin the Fermi energy at the surface, prohibiting inversion or otherwise affecting the device characteristics.

The possibility of thermally grown SiO_2 on Si single crystals as a gate insulator was proposed by Atalla *et al.* [10] in 1959 and good-quality SiO_2 films were first grown by high-pressure steam oxidation of Si by Ligenza and Spitzer [11]. Shortly thereafter the first modern field-effect transistor was reported by Kahng and Atalla [12]. This work demonstrated the potential for the $Si-SiO_2$ system and led to a tremendous research effort to control and characterize the interface states. The material characterization effort continues to this day (see, e.g., Pantclides [13]), with many states yet to be characterized and associated with intrinsic defect configurations or impurities. From a practical point of view the results of this effort have lowered the surface state densities remarkably from the few times $10^{11}/cm^2$ range down to below $10^{10}/cm^2$. Without question the greatest milestones in this materials effort occurred in the mid-sixties when it was discovered that the ionic contaminants could be gettered [14] and that the main ionic contaminant was sodium ions. Once identified, rapid progress toward elimination of Na contaminants was made. While this work was progressing, another important discovery occurred. It was found that by selectively etching through the oxides, exposed regions of the silicon could be doped and *self-aligned* contacts made to the inversion layer [15]. These self-aligned contacts considerably eased the fabrication process.

Such discoveries, along with many smaller contributions, made possible the MOS transistor as used in modern-day devices, but they also made the inversion layer a quality device for the studies of a two-dimensional electronic system. Back in 1957, Schrieffer [16] pointed out that band bending in the inversion layer is on the scale of the de Broglie wavelength of the electrons of interest at low temperatures so that their motion perpendicular to the surface would be quantized into discrete levels. It was the mid- to late-sixties when experiments began that explored these devices with this fact in mind [17].

There are a number of reasons for studying a two-dimensional electronic system. From the device point of view, undoubtedly the most important is the basic understanding of high- and low-field mobilities of electrons moving through the channel. Indeed, much of the early work concentrated on measurements of the mobility [17]. From the more fundamental point of view, it is the ability to control the electron density continuously by varying the gate voltage that has made the inversion layer so important. For example, as the electron density decreases, the coulomb interaction between electrons becomes relatively more important than their kinetic energy, and at sufficiently low densities a Wigner transition to a lattice should occur. Although this transition has not been observed, a number of precursor effects have been observed [18] and the transition itself may have recently been observed in an applied magnetic field [19]. Because of the variations in potential in the inversion layer, it has been possible to study the phenomena of electron localization at low densities [18,20]. In two dimensions, the resistance per square can be written in terms of fundamental constants times a dimensionless conductivity that characterizes the system:

$$\sigma = (e^2/\hbar \; \bar{\sigma} = (2 \times 10^{-4}/\Omega)\bar{\sigma}.$$

Here $\bar{\sigma}$ is the dimensionless conductivity. In the simplest case, $\bar{\sigma}$ is proportional to $k_F l$, the product of the Fermi wave vector and the electron mean free path. When this product is of order one, $\bar{\sigma}$ is ~ 0.1 and this has been proposed as the minimum possible metallic conductivity. Conductivities lower than this would necessarily indicate insulating behavior versus temperature. Recently, this concept has been challenged [21] and the question of whether all electron states in two dimensions are localized has been raised. This problem is one of the more fundamental questions in condensed matter physics at the present time.

Other aspects of inversion layers that are of fundamental interest are the magnetoconductivity and the energy band structures particularly as a function of surface orientation. Magnetic fields normal to the surface of the inversion layer cause the band structure to break up into discrete levels which can be filled or emptied by varying the electron density. They give rise to large variations of resistance, cyclotron resonance, and other phenomena. Because of the six valleys in the conduction band of silicon, a complex energy-level structure occurs in the inversion layer [22] and various splittings have been observed as a function of the relation between the surface direction and the crystal axis [23].

Besides the electronic properties of inversion layers, there continues to be considerable interest in the atomic structure of this remarkable interface. Its structure has been examined by electron microscopy, ion scattering, and various other techniques [13], and it has been shown to be remarkably sharp, indeed changing from pure silicon to silicon dioxide in two or less atomic layers. This sharp transition occurs even though there is a large difference between the lattice constants and structures of Si and SiO_2. There have been numerous attempts to fabricate insulating layers with similar properties on Ge, GaAs, and other semiconductors and none have led to the quality of the $Si-SiO_2$ interface.

Although the current interest in the inversion layer from the device and the basic research viewpoints has the questions of defects and mobility in common, the preceding discussion leads naturally to the question of what application will come from the basic research of the seventies? Will the studies of localized states find applications in the future? Perhaps it is too early to tell but they may become important in charge storage or relaxation mechanisms when very high speeds become important. One such application has been found for amorphous silicon where rapid transfer to localized states of conducting electrons can enhance the switching times of picosecond optical switches [24]. Another possibility is that of Block oscillators based on the band structure in the inversion layer as a function of crystal orientation [25]. The desire to study the detailed characteristics of the devices often induces further research to improve the fabrication techniques. As devices continue to shrink, questions regarding the structure of amorphous materials and more detailed information on the interface will be important for improving the dielectric breakdown properties, further reducing the interface state density, and improving the electron mobilities. It is the difficulty in making such predictions that necessitates a broadly based research effort.

B. Ion–Solid Interactions—Ion Implantation

Perhaps one of the more interesting histories regarding the interplay between basic research and the development of integrated circuits is that of ion–solid interactions and their applications, most importantly to ion implantation, but also to backscattering analyses of thin films and impurity location and surface analysis using channeling. The development of ion implantation involved research in nuclear physics as well as the conventional fields related to electronics: solid-state physics and materials science. It is always difficult to define the beginning of research that has led to a given application. Perhaps the subject of ion–solid interactions can be said to have started with Rutherford's famous experiment on the scattering of alpha particles which exhibited nuclear coulomb scattering. However, the first strong impetus for its development came from the need to understand radiation damage effects in reactors starting in World War II. Indeed, many important aspects of the motion of an ion in a solid, including electronic stopping power, nuclear collision rates, were established by the early 1950s [26].

During the immediate post-war period, solid-state electronics was being developed and radiation effects on semiconductors were investigated, particularly at Purdue University and Oak Ridge National Laboratory. Throughout this period the effects of high-energy ions on solids were thought of in terms of "radiation damage" and, as one so often finds in science, the development of ion implantation and the use of ion beams for analysis depended on the reversal of this commonly held view of the subject. It appears that the first person to look for improvement in semiconductor devices from radiation was Ohl [27], who bombarded point-contact silicon diodes with helium ions and observed an improvement in the reverse current characteristics. This improvement was later recognized as being due to an increased surface resistance due to damage and not a doping effect.

Soon afterward, Cassins [28], at Cambridge University, bombarded germanium crystals with a variety of ions and showed that the surface acceptor concentration was increased. His experiments, therefore, represented the first indication that semiconductor could be doped by implantation. The possibility of doping semiconductors via ion implantation was recognized in a series of patents [29] in the late fifties, while the first publication that clearly distinguished between doping and damage appeared in 1961 [30].

During this same period, the idea of "crystal detectors" of energetic particles, that is solid-state diodes that measured the ionizing radiation of a high-energy ion, was established. The solid-state particle detectors, however, did not catch on until the late fifties and early sixties, and it is in the beginning sixties that a steady increase of activity and cross fertilization between the study of ion–solid interactions and electronics development began in earnest. The work at Purdue University on solid-state particle detectors by Mayer and Gassick [31] gave an indication of the possibilities for solid-state detectors and created a strong increase in interest in their development. In 1962, Alvager and Hansen [32] showed that a shallow, large-area, $p-n$ junction suitable for particle detectors could be fabricated by phosphorus implantation into silicon after annealing at 600°C. The implanted phosphorus, after annealing the damage created during implantation, acted as an n-type dopant similar to conventional diffusion doping.

By the early sixties, the development of semiconductor devices had progressed to the point that the control of the dopant purity, concentration, and location was becoming difficult using standard chemical techniques and the need for alternative processes was becoming apparent. Thus the combined interest of the electronics industries and various national laboratories interested in particle detectors drove the studies of ion implantation in the sixties and early seventies. There began a rising interest in exploring the properties of materials doped by implantation and subsequently thermally annealed and in studying the interplay between the need to reduce the damage created by implantation and to electrically activate the dopants. The commercial utilization of ion implantation in semiconductors is widespread today. Interestingly, almost all applications make use only of the advantages implantation offers in concentration and impurity control, the spatial distribution being established by diffusion.

The study of ion implantation was greatly enhanced in the mid-sixties by the discovery of ion channeling in crystalline solids. In theoretical simulations [33] of the motion of energetic ions in crystalline solids, it was found that there was an anomalous penetration of ions when the beam was directed parallel to the symmetry directions of the crystal. Indeed, it was soon realized that the crystalline potential actually guided the ions in the space between atoms. Thus the ion is said to be channeled by the rows of atoms. The phenomena of channeling was extensively studied in the sixties and was quickly recognized as an important technique for analyzing the effects of ion implantation. If, for example, an implanted atom is in a substitutional site in a lattice, the channeled beam cannot scatter from it any more than from the host crystal atoms, but if it is sitting in the channel between crystal rows, it will scatter the channeled beam more strongly than a beam from a random direction. Thus the dopant lattice position could be established and the detailed effects of annealing could be studied.

In the early seventies, as planar technology began to develop in earnest, the use of Rutherford backscattering for purposes of chemical analysis became an important contribution to yet another aspect of microfabrication. As a high-energy (megaelectron volt) ion enters a solid, it primarily loses energy to collisions with electrons. These inelastic collisions do not deflect the ion but cause it to gradually lose energy. It is the relatively infrequent hard collisions with nuclei that backscatter the ion. When backscattered out of a solid, the ion's energy is given by the total distance it traveled while losing energy to the electrons in the solid plus the energy lost in the nuclear collision. Since both these processes are well understood, the energy spectrum of the backscattered beam can be used to profile the atomic constituents of thin films. With the need for metallic leads in planar technology came the need to understand interdiffusion between different thin films, and there has been extensive use of backscattering for thin-film analysis [34]. Probably one of the more interesting and technologically important systems is that of the transition-metal silicides. These films are widely used as conducting paths on integrated circuits. In the particular case of Pt on silicon, it was found that even at low temperatures a solid-state reaction occurs between the silicon and the Pt and that this reaction takes place through well-defined intermediate phases. The first phase to grow was Pt_2Si, which was then followed by PtSi as the Pt layer is depleted. The contact metallization using silicides was developed by Lepselter working on integrated circuit processing [35], but the discovery of the low-temperature solid-state reactions has lead to major research programs to understand the chemistry of these solid-state reactions in terms of the equilibrium phase diagram of the materials. This research has now come to the point that extremely well characterized epitaxial nickel silicide films can be grown on silicon [36]. Such films may have device potential in the future.

Ions have yet another important application in the analysis of surfaces and interfaces. As microfabrication continues to decrease the dimension of devices, surfaces and interfaces become of increasing importance. Thus in the past few years a variety of techniques has been used to analyze the nature of the $Si - SiO_2$ interface, the $Si - MSi$ interface, and the surface structure of clean silicon on an atomic scale.

On another front, scientists interested in developing high-density plasmas for fusion purposes have developed a new type of ion source— a liquid-metal source that has a high brightness—and work at Hughes [37] has shown that this source can be used to sputter lines in Al with a width of 450 Å. These sources are made by using a solid tip of a few microns in radius projecting from the liquid metal. The liquid flows up the tip and forms a sharp cone which acts as the bright source. Currently, increased research efforts toward a better understanding of the operating characteristics of these sources are underway. If improvements are found, ion beams generated in this way may find applications in the future.

In the past few years, it has been found that lasers and electron beams can be used to locally heat the implanted layer to remove the damage from ion implantation. Thus laser annealing offers the potential of replacing the standard thermal anneal. The local heating has the advantage of avoiding possible side effects of thermal anneals on other parts of an integrated circuit. It has been shown that, depending on the intensity and pulse length, there are two ways in which a laser beam can cause annealing of an implanted layer (see, e.g., Ferris et al. [38]). For nanosecond pulses of megawatt powers, the surface layer is found to be heated very rapidly and melts to depths of several thousand angstroms. As it cools, the crystal regrows epitaxially from the substrate. On the other hand, for lower-power (kilowatts) continuous laser exposures, one obtains solid-state epitaxial regrowth. These two regimes are characterized by the differences in dopant diffusion and various other properties. Currently, one of the most interesting possible applications of laser annealing is the crystallization of amorphous silicon on dielectric substrates. Large-grain polysilicon has been grown by this technique.

The more recent developments in ion implantation and the uses of ion beams cited above illustrate the diversity of areas of science and technology that, when brought together, represent *scientific* and *technological* progress. The phenomenon of laser annealing is allowing us to study the kinetics of fast growth and the creation of metastable alloys. At the same time, it offers the possibility of a new technological processing.

Laser annealing, in turn, is made possible by the tremendous developments in lasers, ion implantation, and semiconductor materials processing.

The development of the silicides started in semiconductor device work but has strongly impacted our understanding of solid-state chemistry and is of interest for interface studies. In turn this knowledge of the silicides forms a base for further use of silicides in integrated circuits. Ion implantation originates from fundamental studies of ion–solid interactions, semiconductor research, and nuclear physics. Its development has led to practical applications, which have been used to further develop solid-state detectors for nuclear physics. Thus we encounter, again and again, the necessity of research and development on a broad front with interactions and communications across these boundaries in order that progress occurs in all of science and technology.

III. MICROFABRICATION AND BASIC RESEARCH TODAY

As in the two examples just discussed, microfabrication is coupled to present-day research in materials science, physics, and chemistry in a variety of ways. The purpose of this section is to mention a few examples of these interactions. In materials science, microfabrication continually challenges our fabrication ability, while in physics, it often offers a new technique or device for doing experiments. In addition, it is one of several general motivating forces for research on surfaces and interfaces.

A. Materials Science

Materials synthesis, purification, and doping, thin-film deposition, resist control, and materials sensitivity to radiation, along with the many other aspects of materials processing, set the current limits on microfabrication. It is quite simple for the physicist or engineer to argue that current integrated circuits can be scaled down to the few tenths of a micron level, but it is quite another matter to accomplish this scaling because it requires improvements in the quality of substrates, deposited films, masks, resists, and control of the interactions among various films, resists, and implants. Consequently, materials science in many ways sets the pace for microfabrication. (For a more complete discussion of materials science related to microfabrication, see other chapters in this volume.) Here I would like to mention briefly a few of the current challenges to research in materials science and a few directions in materials research that may impact in future microstructure science.

1. Challenges in Materials Research

Much current interest centers on assessing the potential of III–V compounds as alternatives to silicon-based integrated circuitry and on the further development of low-temperature processes for silicon.

A major objective of recent research has been the exploration of GaAs-based integrated circuits. It is believed that such circuitry has the potential for higher speeds, primarily because of the higher saturation velocities. In addition, one may find advantages in coupled optical and electronic integrated circuitry entirely fabricated in GaAs. Devices in GaAs are based on Schottky barriers operated in the depletion mode, and much progress has been made. The challenge for research has been to fabricate good dielectric layers on GaAs so that inversion-layer-type devices can be realized. In this work, efforts have concentrated on GaAs oxides, and various growth techniques have been attempted [39]. The basic difficulty is that, of the two constituents of the oxide, As_2O_3 is relatively unstable and, in chemical equilibrium with Ga, will decompose into As + Ga_2O_3. Thus one generally finds that metallic clusters of arsenic form at the interface or in the bulk oxide and destroy the dielectric characteristics. It is interesting that the interface between the oxide and GaAs is found to be spatially very sharp, particularly for anodic-grown oxides kept at low temperatures. Various techniques have been tried for the purpose of eliminating the metallic As. Perhaps the most successful [40] has been the use of fluorine in plasma growth where interface state densities of 2 to 3 × 10^{11}/cm^2 have been obtained. At this stage it is difficult to predict whether we are in the late fifties relative to the $Si–SiO_2$ story or whether GaAs is the germanium of the III–V compounds as far as dielectric layers are concerned.

Perhaps III–V technology will not use MOS-type structures and an interesting possible means of fabricating devices has been exhibited in GaAs using molecular-beam epitaxy. The molecular-beam epitaxy (MBE) adds a versatility in GaAs that may ultimately be this material's real strength. To illustrate this, unipolar devices have been fabricated [41] in which a Schottky-like barrier is created by grading the composition from GaAs to AlAs on the scale of a few hundred angstroms. In this way one obtains a barrier of 0.4 eV, which acts as a simple diode. Since this is contained in the Ga–AlAs, no dielectric layers are involved.

Although much work is clearly needed in GaAl devices, the latest trends in optical communications has been toward the use of 1.3-μm radiation and this trend raises the question of fabricating devices in other III–V compounds. Such work is just beginning.

In silicon-based technology, there continue to be challenging materials problems. One is the growth of thin, large-grain or single-crystal silicon on dielectric substrates. Realization of a fabrication technique of such a film is important for several reasons. First, as MOSFET devices have become smaller, the parasitic capacitances that occur between the source or drain and the substrate, for example, have become the limiting factors controlling speed. Such problems could be removed by using thin silicon on a dielectric. Second, in both large-area display technology and in solar cell technology, large-grain or single-crystal silicon grown by an inexpensive process on a cheap substrate is essential. Recently, Gat et al. [42] have shown that laser annealing of polysilicon can change its grain size by over an order of magnitude and some success has been obtained by Geis et al. [43] using a technique called graphoepitaxy. Other techniques are being attempted and continued progress can be expected in the future.

As device dimensions become smaller, materials control becomes increasingly important and low-temperature processing receives more emphasis. Generally, one would like low-temperature processes in order to avoid the deleterious effects of high-temperature diffusion, defect creation, etc., on one part of a circuit while fabricating another part. In the past few years considerable effort on the basic physics side has been placed on understanding the growth of silicides and on laser annealing of implantation-doped silicon. Both of these have been mentioned in Section II.B on ion–solid interactions. Because local temperatures are high, the technique tends to create point defects that must be eliminated to make useful devices. Several schemes for doing so have now been exhibited [44].

Plasma growth of thick SiO_2 at low temperatures is being explored at several laboratories and could be very worthwhile. The currently used technique is a high-pressure steam growth at 700 to 900°C. Plasma techniques in general, for example, etching, and deposition, are just beginning to be understood from a fundamental point of view and much work is left to be done.

These are but a few examples of materials problems in microfabrication. Those working in the field can surely make a much longer list. Again the technologically important subjects mentioned here are having an impact on our fundamental understanding of materials science. Studies of silicon dioxide are continuing to allow us to learn more about the structure of amorphous materials. Such questions as the nature of microvoids in amorphous materials are being investigated in SiO_2. Laser annealing is revealing the limiting times for recrystallization and the differences in melting temperatures of amorphous and crystalline materials and may lead to new insights regarding instabilities in crystal growth. This interplay continues in much of materials sciences.

B. Physics Research

Physics research, particularly condensed-matter physics, is coupled to microfabrication in a variety of ways. The number of experiments that depend on microfabrication techniques is increasing, while at the same time much of condensed-matter research has long-term implications on microfabrication. In this section we shall give several examples in the former category and shall then discuss progress in surface and interface science as an example of the latter. A new set of tools for characterizing surface structures and properties are being developed which are making possible new advances in the understanding of surfaces and interfaces. The long-term implications of this development will be discussed in the last sub-section.

1. Microfabrication Related to Experiments

As discussed in Section II.A. on inversion layers, the question of localization of electron wave functions due to potential energy fluctuations induced by disorder is fundamental to solid-state physics. Recently, Thouless [45] predicted that any one-dimensional wire whose cross section is sufficiently small relative to its length would become insulating at low enough temperatures. This prediction, along with the questions regarding localization in two dimensions, have created great theoretical and experimental interest. Wires as thin as several hundred angstroms have been fabricated to obtain one-dimensional behavior while relatively thin films have exhibited two-dimensional behavior and a new logarithmic behavior of the resistivity at low temperatures has been observed [46]. This research is testing our abilities to fabricate extremely narrow metallic wires.

In high-pressure physics it has been found that very high pressures can be attained in the laboratory provided the pressure is concentrated on extremely small micron-sized samples. In order to make measurements of the electrical properties of such samples, it is necessary to fabricate a set of interdigitated electrodes with leads separated by ~0.1 μm. Such a technique was recently used to show that solid neon can be made metallic at high pressures [47]. This technique is being picked up at other laboratories.

As one reduces the size of materials in a controlled way, a number of lengths, characteristic of various states of matter, become accessible. The most obvious of these is the superconducting coherence length, but there are others such as the quasi-particle diffusion length in a superconductor, the electron mean free path, the size of domains in incommensurate charge density waves, and the coherence length of the charge density wave itself to name a few. Thus direct measurements of these lengths, such as the measurement of the quasi-particle diffusion length by Dolan and Jackel [48], become feasible. In a charge density wave system very little is known about the structure of domain walls or the distribution and magnitude of the charge carried by such walls. Perhaps in the future these properties will be measured by microfabricated devices.

Silicon inversion layers are in some sense a two-dimensional waveguide for electrons. That is, the inversion-layer thickness is of the order of the de Broglie wavelength of thermal electrons. Perhaps good epitaxial metals will allow creation of one-dimensional waveguides, i.e., quantized levels in two of three dimensions.

A possible use of microfabrication may be in what might be called pinhole spectroscopy. Suppose one could purposely introduce controlled small holes approximately 100 Å in diameter in an oxide between two metals. Then, with an applied voltage, electrons could be emitted through the hole with considerable energy, thus coupling to phonons of the bulk materials of either side. Such experiments have been performed with random pinholes by Yanson and Kulik in the Soviet Union [49]. Perhaps this technique could be made into a more precise spectroscopy using microfabrication.

Far-infrared spectroscopy has been limited by a lack of detectors, mixers, and local oscillators. Photodetectors cease to be available below a few microns in wavelength (10^{14} Hz), while diodes normally operate below 100 GHz in frequency. The latter are limited by the RC of the device. Since the capacitance is related to the cross-sectional area, microfabrication can be used to increase the accessible frequency. Indeed, GaAs–Au Schottky diodes with 1-μm dimensions have been shown to operate near 10^{11} Hz and 0.4-μm^2 superconducting tunnel junctions have been shown to work as mixers at 110 GHz [50,51]. The latter devices have been used to make the first observations of carbon in interstellar clouds. Additional work in further reducing the size of these devices will lead to even higher frequencies. In addition to this direct application of microfabrication to astrophysics, the development of CCD imaging arrays is having a major impact on visible spectroscopy because their high resolution and dynamic range enhance the ability to pick out point objects.

In x-ray research, microfabrication is being used to make Fresnel zone plates. These plates test our microfabrication abilities to the limit as they require circular lines with widths of 0.1 μm and very high aspect ratios. These are being developed for studies of laser fusion [52] but may eventually be useful as a focusing element. Yet another use of microfabrication may be in x-ray microscopy, where one uses a single pinhole array to perform scanning x-ray microscopy. One would like to make the hole as small as possible, as the hole determines the resolution of the technique.

As it becomes possible to work in the submicron range and the microfabrication techniques become more accessible to basic research scientists, many other uses will be found. Microfabrication simply allows one to set up experimental measurements on the micron scale or, for research purposes, submicron scale. Although it is not the purpose of the chapter to discuss the overall impact of integrated circuit technology on research, the restricted discussion given here based on microfabricating devices for experiments has omitted another revolution taking place in the way in which physics research is carried out. Microprocessors and minicomputers are standard equipment in most experiments today and are allowing data acquisition and on-line analysis and decision making that have revolutionized research.

2. Surfaces, Interfaces, and Two-Dimensional Systems

As emphasized in other chapters of this book, surfaces and interfaces are becoming increasingly important in microfabrication. In fact, along with this increasing importance, surfaces, interfaces, and other two-dimensional systems have become the mainstream of condensed-matter physics research. This research has led to gradual improvement in surface characterization and to the development of various ultraviolet spectroscopies, which have led to an enormous increase in our understanding of the electronic spectra of surfaces. In addition, both ion-beam and x-ray techniques are being developed to complement conventional low-energy electron diffraction (LEED) to determine surface structures and phase transitions.

The progress on the electronic front has been most impressive. A variety of spectroscopies, photoemission, particularly angle-resolved photoemission, photoelectron diffraction, core-level spectroscopy, and elipsometry have been developed. Without doubt, the most important of these has been angle-resolved photoemission [53]. It has turned out to be possible, by measuring the angle at which the photoelectron is emitted, to choose electron states of definite momentum parallel to the surface and, using energy resolution, to identify direct transitions between energy bands along the perpendicular momentum direction. This has allowed the mapping out of both surface and volume energy bands. At the same time there has been tremendous progress on the theoretical front toward calculating surface electronic states. These calculations have allowed direct comparisons of the theoretical photoemission spectra with experiments and thus an indirect determination of surface relaxation and reconstruction [54].

It has taken some forty years to go from the first electron diffraction experiments of Davison-Germer to the present level of understanding of surface structures. Although we understand many aspects of the reconstruction of surfaces and interfaces, we still do not have good, reliable ways of determining the positions of atoms, particularly with respect to their perpendicular position. Much effort has gone into the calculation of LEED intensities as a function of energy, but the problems of energy loss and multiple scattering have plagued these calculations and have led to definitive structures in only a few cases.

In the past several years, ion scattering has been used successfully to complement LEED work. By using various aspects of channeling surface peak structures and shadowing, direct information regarding atomic positions can be obtained [55]. In addition, low-energy ion scattering can be used to determine surface concentrations and positions by energy loss and shadowing considerations [56].

More precise information appears to be coming from several x-ray techniques applied to surfaces. Extended x-ray absorption fine structures, which can be interpreted to give distance to atoms surrounding specific atoms, have been applied to a variety of overlayers and have been shown to give precise information about the local environment, both in terms of distances and coordination [57].

Another interesting x-ray technique is the use of standing waves that exist at the surfaces of perfect crystals without mosaic spread [58]. The standing-wave field is established on the atomic scale, and its node can be adjusted relative to the atomic position in the bulk of the sample. This technqiue has been applied to Br in organic solution above the surface of silicon and has been shown to give accurate information on the Br–Si distances.

One of the more promising x-ray techniques for determining detailed structures of ordered overlayers is the reflection–diffraction technique [59]. In the glancing-angle total-reflection mode, the x-ray intensity is enhanced near the surface and Bragg diffraction from reconstructed surfaces can be obtained. Because there is no problem with multiple scattering, the intensities of the Bragg reflections should be interpretable in the same fashion as in conventional x-ray crystallography. This technique, as well as the standing-wave technqiue, can be used on interfaces.

The reason for bringing up this progress here is to point out that the progress in this area is not dissimilar to that which occurred in the fifties in bulk crystals. As will be discussed in the final section, perhaps controlled epitaxial interfaces will be used as the thin film of the future and its particular electronic structure will be manipulated for device purposes.

IV. FUTURE DIRECTIONS OF BASIC RESEARCH

It is quite clear that microfabrication techniques will move into the tenths of a micron range in the next ten years. As this happens, further reduction will be challenged by a number of "fundamental limitations" inherent in scaling present-day devices. These have been discussed by various authors [60]. Basically, the reason for going to smaller dimensions is to reduce the product of the switching time times the power. Since this product is dependent on such quantities as the load capacitance, one can reduce its value by reducing the dimensions.

Below ~1000 Å electric fields, particularly in the depleted regions of the semiconductor, can become large enough to cause breakdown in devices operating with voltages greater than the bandgap voltages. In addition, the fluctuations in dopant concentrations become a serious statistical problem. Other effects such as hot-carrier penetration of dielectric layers and electromigration become more important. For these reasons, many feel that the current microfabrication technology will stabilize with circuits with the smallest features on the scale of a few thousand angstroms.

It is clear that overcoming this limiting region is microfabrication's challenge to the basic research of the future. Will the present-day efforts in surface and interface physics discussed briefly in the previous section give rise to new technology to overcome some of those limitations to scaling? It is likely that they can. The limitations of dopant fluctuations in one example that modern surface science could perhaps eliminate. If one could dope a semiconductor using a crystalline ordered overlayer, then the fact that it is ordered eliminates the random statistics of conventional semiconductor doping. In fact, graphite can be doped by ordered layers of alkali metals and other compounds. Perhaps microfabrication can be done by electron-beam desorption and definition of overlayer regions of a surface.

Can we find larger units of charge than the electron charge so that the nonlinearity of devices is increased? Perhaps a domain wall in a two-dimensional charge-density wave could ultimately be used as a unit of charge.

To some readers the previous suggestions sound a bit like science fiction. However, these suggestions are put forth in order to challenge the imagination of those in basic research regarding the ultimate ways of fabricating the microcircuits of the future.

REFERENCES

1. W. Shockley, "Electrons and Holes in Semiconductors." Van Nostrand-Reinhold, Princeton, New Jersey, 1950.
2. D. Kahng, *IEEE Trans. Electron Devices* **ED23**, 655 (1976).
3. S. M. Sze, "Physics of Semiconductor Devices." Wiley(Interscience), New York, 1969.
4. M. Pepper, *Contemp. Phys.* **18**, 423 (1977).
5. J. E. Lilienfeld, U.S. Patents 1,745,175 (1930); 1,900,018 (1933).
6. O. Heil, British Patent 439457 (1935).
7. S. Shockley and G. L. Pearson, *Phys. Rev.* **74**, 232 (1948).
8. W. L. Brown, *Phys. Rev.* **91**, 518 (1953).
9. I. M. Ross, U. S. Patent 2,791,760 (1957).
10. M. M. Atalla, E. Tannenbaum, and E. J. Scheibner, *Bell Syst. Tech. J.* **38**, 749 (1959).
11. J. R. Ligenza and W. G. Spitzer, *J. Phys. Chem. Solids* **14**, 131 (1960).
12. D. Kahng and M. M. Atalla, *IRE-AIEE Solid-State Device Res. Conf.* Carnegie Institute of Technology, Pittsburgh, Pennsylvania, 1960.
13. S. T. Pantelides, ed. "The Physics of SiO$_2$ and Its Interfaces". Pergamon, Oxford, 1976.
14. D. R. Kerr and D. R. Young, U.S. Patent 3,303,059 (1967).
15. J. C. Sarace *et al., Solid-state Electron.* **11**, 653 (1968).
16. J. R. Schrieffer *in* "Semiconductor Surface Physics" (R. H. Kingston, ed.), p. 55. Univ. of Pennsylvania Press, Philadelphia, Pennsylvania, 1957.
17. F. F. Fang and A. B. Fowler, *Phys. Rev.* **169**, 619 (1968).
18. F. Stern and W. E. Howard, *Phys. Rev.* **163**, 816 (1967).
19. B. A. Wilson, S. J. Allen, Jr., and D. C. Tsui, *Phys. Rev. Lett.* **44**, 479 (1980).
20. N. F. Mott, *Phil. Mag.* **26**, 1015 (1972).
21. E. Abrahams, P. W. Anderson, D. C. Licciardello, and T. V. Ramakrishnan, *Phys. Rev. Lett.* **42**, 673 (1979).
22. L. J. Sham, S. J. Allen, Jr., A. Kamgar, and D. C. Tsui, *Phys. Rev. Lett.* **40**, 472 (1978).
23. D. C. Tsui, D. M. Sturge, A. Kamgar, and S. J. Allen, Jr., *Phys. Rev. Lett.* **40**, 1667 (1978).
24. D. H. Auston, P. Lavalland, N. Sol, and D. Kaplan, *Appl. Phys. Lett.* **36**, 66 (1980).
25. L. Esaki and R. Tsu, *IBM J. Res. Dev.* **14**, 61 (1970).
26. N. Bohr, *Kgl. Danske Vid. Selsk. Matt-Fys. Medd.* **18**, No. 8.
27. R. Ohl, *Bell Syst. Tech. J.* **31**, 104 (1952).
28. W. D. Cussins, *Proc. Phys. Soc. London* **B68**, 213 (1955).
29. R. Ohl, U.S. Patent 2,750,541; W. Shockley, U.S. Patent 2,787,564.
30. F. M. Rourke, J. C. Sheffield, and F. A. White, *Rev. Sci. Instrum.* **32**, 455 (1961).
31. J. W. Mayter, *J. Appl. Phys.* **30**, 1939 (1959).
32. T. Alvager and N. J. Hansen, *Rev. Sci. Instrum.* **33**, 567 (1962).
33. O. S. Oen and M. T. Robinson, *Appl. Phys. Lett.* **2**, 83 (1963).
34. J. M. Poate and King-Ning Tu, *Phys. Today* **33**, 34 (1980).
35. M. P. Lepselter, *Bell Syst. Tech. J.* **40**, 233 (1966).
36. K. C. R. Chiu, J. M. Poate, L. C. Feldman, and C. J. Doherty, *Appl. Phys. Lett.* **36**, 544 (1980).

37. R. L. Siliger, J. W. Ward, V. Wang, and R. L. Kobena, *Appl. Phys. Lett.* **34,** 310 (1979).
38. S. D. Ferris, H. J. Leamy, and J. M. Poate (eds), Laser-solid interactions and laser annealing, *AIP Conf. Proc. #50* (1979).
39. B. Schwartz, *Crit. Rev. Solid State Sci.* **5,** 609 (1975); R. P. H. Chang, *Thin Solid Films* **56,** 89 (1979).
40. R. P. H. Chang, J. J. Coleman, A. J. Polak, L. C. Feldman, and C. C. Chang, *Appl. Phys. Lett.* **34,** 237 (1979).
41. C. L. Allyn, A. C. Gossard, and W. Wiegmann, *Appl. Phys. Lett.* **36,** 373 (1980).
42. A. Gat, L. Gerzberg, J. F. Gibbons, T. J. Magu, J. Peng, and J. D. Hang, *Appl. Phys. Lett.* **33,** 775 (1978).
43. J. M. W. Geis, D. C. Flanders, and H. L. Smith, *Appl. Phys. Lett.* **35,** 71 (1979).
44. J. L. Benton, C. J. Doherty, S. D. Ferris, D. L. Flamm, L. C. Kimerling, and H. J. Leamy, *in* "Laser and Electron Beam Processing of Materials," (C. W. White and P. S. Peercy, eds.) Academic Press, New York, 1980. p. 430.
45. D. J. Thouless, *Phys. Rev. Lett.* **39,** 1167 (1977).
46. G. J. Dolan and D. D. Osheroff, *Phys. Rev. Lett.* **43,** 721 (1979).
47. A. Ruoff. *Phys. Rev. Lett.* **42,** 383 (1979).
48. G. J. Dolan and L. C. Jackel, *Phys. Rev. Lett.* **39,** 1628 (1977).
49. I. K. Yanson and I. O. Kulik, *Proc. Int. Conf. Low Temp. Phys., 15th, Grenoble, France* **3** (1971).
50. A. Y. Cho, J. V. DiLorenzo, B. S. Hewitt, W. C. Niehaus, and W. O. Schlosser, *J. Appl. Phys.* **48,** 346 (1977).
51. P. L. Richards, T. M. Shen, R. F. Harris, and F. L. Lloyd, *Appl. Phys. Lett.* **34,** 345 (1979); G. J. Dolan, T. G. Phillips, and D. P. Woody, *ibid.* **34,** 347 (1979).
52. N. M. Ceglio, H. I. Smith, *Rev. Sci. Instrum.* **49,** 15 (1978).
53. J. A. Appelbaum and D. R. Hamann, *Surf. Sci.* **74,** 21 (1978).
54. D. E. Eastman, *J. Vac. Sci. Technol.* **17,** 492 (1980), and references therein.
55. L. C. Feldman, "Surface Science: Recent Progress and Perspectives" (R. Vanselaw, ed.). C. R. C. Press, Cleveland, Ohio, 1980.
56. H. H. Brongersma and T. M. Buck, *Nucl. Instrum. Methods* **148,** 569 (1978).
57. P. H. Citrin, P. Eisenberger, and R. C. Hewitt, *Phys. Rev. Lett.* **41,** 309 (1978).
58. P. L. Cowan, J. A. Golovchenko, and M. F. Robbins, *Phys. Rev. Lett.* **44,** 1680 (1980).
59. W. C. Marra, P. Eisenberger, and A. Y. Cho, *J. Appl. Phys.* **50,** 1627 (1979).
60. R. W. Keyes, *Proc. IEEE* **63,** 740 (1975); see also B. Gopinath and G. L. Miller, unpublished preprint.

Chapter **6**

Impact of VLSI on Medicine

JACOB KLINE

Department of Biomedical Engineering
University of Miami, Coral Gables, Florida

I. INTRODUCTION

With its low cost, small size, and great versatility, VLSI technology is destined to have a most significant impact on medicine and the delivery of health care. Cost containment will occur via increased clinical and administrative efficiency through better facilities for hospital management and medical research. VLSI will improve the cost effectiveness of medical devices through advanced, smaller-sized circuitry, which provides increased sophistication and enhanced function. It will make possible the

171

development of implantable devices with closed-loop, regulated control to artificially substitute for the functions of the pancreas, liver, kidney, and other organs or glands.

Other chapters in this book will treat microstructured components, microprocessor device specifications, and performance characteristics, as well as production techniques. This chapter will present a cross section of the most significant advances that will be made through VLSI in the decade of the 1980s, which will revolutionize the delivery of health care and improve medical practices and research. The first section will treat advances in the area of the delivery of health care in the hosptial, clinic, and private practice; elements of patient record keeping, patient monitoring systems, computer usage in the clinical laboratory, diagnostic process, and patient management are addressed. The remaining sections treat computerized scanning and nuclear techniques, prosthetic and orthotic devices, hemodialysis, and, finally, computerized biofeedback.

This chapter does not cover, individually or specifically, the advances, costs, size reductions, and extended capabilities that will be experienced by conventional devices that have been in existence for the past two decades. These include patient monitoring instrumentation, such as the electrocardiogram, blood pressure and blood flow devices, electrosurgery instrumentation, defibrillators, anesthesia apparatus, implantable transducers, and biotelemetry systems. However, new generations of these devices will undoubtedly be developed with VLSI components and their impact will be implicit in the systems presented here. New generations of the implantable cardiac pacemaker, which is a device with a long successful history, will be treated in the section on prosthetic devices.

In the past decade, knowledge of how to use computers in the health care field has rapidly advanced. The problems posed by size, reliability, and the inherent high costs of earlier specialized sophisticated computer systems with discrete electronics severely limited their accessibility to laboratories, clinics, and hospitals. It has been only since the development of microminiature circuitry that the costs and size have been reduced, and various processes have been made available in which theoretical knowledge now may be practically applied.

II. HEALTH CARE DELIVERY SYSTEMS

In the last decade most hospitals have introduced and involved computers in their business affairs, especially in the area of billing. Only in a relatively few of the larger ones has the computer become actively involved in the clinical setting. Some of the major reasons why smaller hospitals have not made greater use of the computer and sophisticated instrumentation are the relatively high cost of the hardware available during the 1970s, the space limitations, and the complex operating systems, which require programmers and dedicated support personnel [1]. In the past, technology has been blamed as the basis for proliferation of health care costs. Although specialized areas to deliver health care, such as coronary care units, intensive care units, and kidney dialysis centers, place expanding demands on Medicare and Medicaid and have increased costs, they definitely have been instrumental in decreasing mortality rates.

The imminence of very large scale integrated circuits applied to medical instrumentation and computers with advanced microstructural components will bring the benefits of low-cost systems closer to the physician and the patient. As a result, more effective delivery of health care to a greater number of the world's population will be possible.

This section will deal with the impact that micro-main-frame computers will have on the transfer of information among all departments in the hospital, the improvement of patient management and diagnostic techniques, and the simplification of finance functions. VLSI technology will increase packing densities (the number of logic elements that may be incorporated on a single substrate) to 200 to 1000 gates/mm², word lengths to 32 to 64 bits, with processor instruction cycle times in excess of 10 MHz. This will enable microminiature-sized computers to approach the characteristics of present, large, expensive main-frame computers. These computer systems will incorporate new generations of random access memory (RAM), read only memory (ROM), programmable read only memory (PROM), and erasable programmable read only memories (EPROM) with capacities of at least 250K bits/chip, resulting in multi-megabyte memories with access times of less than 20 nsec. With gallium arsenide and silicon on sapphire devices, speed–power products on the order of 0.01 to 0.1 pJ will be realized. Bubble memories with capacities of 4 megabits or greater with large external nonvolatile memory systems will be available [2]. High-level software, such as Pascal-like languages, will be routinely incorporated internally into the basic processor instruction set. This will reduce initial software development costs and system development time. These systems will replace expensive, bulky, unreliable, electromechanical storage devices such as hard disks, floppy disks, and tape units [3].

A. Computerized Hospital Organization

Figure 1 shows a block diagram of an advanced hospital organization with a central computer service as an informative control and communication center. Notice that there exists a complex array of departments

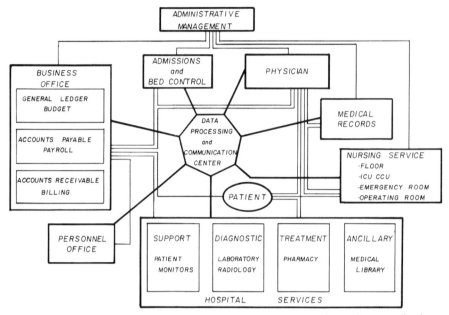

Fig. 1. Hospital organizational chart with central data processing and communication center.

that are strongly interrelated and interconnected and among which there must be close coupling for information transfer and processing. Not only will such nonclinical functions as billing, patient scheduling, and administrative management interact via the computer links, but also the clinical functions such as patient monitoring, laboratory analyses, radiology, evaluations, will be integrated in one complete communication system. In addition, for each component, such as the intensive care unit, a satellite computer system will provide direct, on-line, real-time patient monitoring. Thus will be integrated the diagnostic functions so that at any time ancillary personnel (nurses, physicians' assistants, etc.) can retrieve such information to evaluate the instantaneous status of the patient and also provide enough information to predict the prognosis over the next 24-hr period [4].

Until the advent of the computer, the techniques used for the processing of the information have been extremely rudimentary. Manual systems have suffered from forgetfulness, inaccuracy, and incomplete inclusion of the total body of medical knowledge. Through the use of the computer, new techniques may be provided for the management of information and support investigations previously impossible in the delivery of health care and research.

In the hospital, or the clinic, fully computerized treatment may begin with the obtaining of information pertinent to the patient's history. It is termed *patient data base acquisition*. The patient's record acquisition can also be the basis for the initial input of a computerized, multiphasic, automated, health testing system [5]. From there the computer could be used in a diagnostic support system which would be affected by the acquisition of an analog signal through analog/digital (A/D) conversion and would provide a facility for automatic interpretation. The computer offers diagnostic support in the areas of electrocardiography [6], electroencephalography, pulmonary functions, electromyography, computer axial tomography (CAT scanner), nuclear medicine, and diagnostic consultation through instrumentation. Information from patient support systems, which include the areas of inhalation therapy, occupational therapy, physical therapy, and central stores can also be processed by the central computer. Menu planning, nutritional accounting, food control elements of procurement, inventory preparation, and cost accounting data components may be handled by a computerized food service system. In the clinical laboratory, the computer plays a significant part in the areas of chemistry, cytology, microbiology, immunology, and blood banking.

With VLSI devices it will be possible to develop portable miniature desk-top computer systems to assist and make more effective surgical pathology, tumor specimen analysis, autopsy reporting, forensic pathology, quality control, trend analysis, laboratory instrument monitoring, and positive control of specimens of body fluids and blood and clinical laboratory systems. It has been shown that computers in the intensive care unit, augmenting physiological monitoring, have been effective in controlling and treating trauma. In the radiological area for the analysis of x rays, computers serve image data analysis and pattern-recognition functions.

B. Extension of MUMPS Capability

Software developed by the Massachusetts General Hospital, called the Multi-Utility Multi-Programming System (MUMPS) [7], provides over 300 programs covering medical record keeping, scheduling, intensive care, clinical laboratory data, data management, and other practical applications. The organization, assembly, and compilation of these programs, now economically feasible only for large computer installations, will be transferable to many smaller installations spread throughout the health care facility making use of low-cost, micro-main-frame computers by way of high-capacity VLSI memories on the order of millions of bytes. These remote computers will interact and share information through high-speed communications channels. The emphasis will be on distributed processing with reciprocal sharing of large, central data bases. It will be possible with the new generation of micro-mini-frame computers to utilize the complicated MUMPS applications software, developed primarily for large, main-frame, multiuser operating systems. This will be accomplished at a fraction of the costs that are now inherent in the comparable programming of the present-day conventional computers.

Another software system which is available is the Problem-Oriented Medical Information System (PROMIS). This system correlates and records all data according to the relevant medical problems and reduces the costs and delays of manual entry, provides the physician with a constant guide to the treatment of a particular disease, and advises him of drugs and therapy that are contraindicated. The cost of the application of PROMIS is approximately $5.00 per patient per day. The cost is justified because it eliminates repeated radiology, laboratory, and other tests and minimizes adverse drug reactions. Because the system provides effective preadmittance diagnoses, it can reduce hospital stays and greatly reduce the manpower needed to handle paperwork [8].

With the advent of low-cost micro-main-frame systems with large memory and multisensor capability, software packages previously developed for large systems such as MUMPS and PROMIS can be transported over to micro-main-frame systems with a significant reduction in capital investment. The software systems then could be well within the reach of the budgets of thousands of hospitals. Over several hundred thousand physicians will be able to cut their costs and bring both therapeutic and diagnostic computer system procedures into their environment.

C. Center for Disease Control System

Another significant software system for evaluation of laboratory services has been developed by the Center for Disease Control [9]. With this system the average cost per minute for hematology tests has been considerably reduced. With increased memory capacity and processor capability throughout, realized through the use of VLSI and inexpensive external memory systems [10], fixed, hard disks and multimegabyte bubble memories, smaller laboratories will have access to this evaluation system and, consequently, will reduce considerably the cost of blood profiles, which in the United States alone, amounts to hundreds of millions of dollars per year.

Costs, primarily, and size have been serious limitations to the incorporation of computers into the many areas described. However, with the advent of VLSI, both costs and size of electronic devices will be considerably reduced; greater sophistication will also be added. A greater number of clinics and hospitals will, therefore, be able to introduce instrumentation, the capability of which will be vital to their needs, resulting in more effective and efficient health care delivery at reduced costs.

III. COMPUTERIZED SCANNING AND NUCLEAR X-RAY TECHNIQUES

One of the most significant advances in medical diagnosis in the past decade has been the development of the computerized axial tomographic scanner (CAT) [11]. The Nobel Prize in Medicine in 1979 was awarded to the inventor of this system, G. N. Hounsfield, an electrical engineer. The CAT scanners have revolutionized noninvasive diagnostic processes for dynamic organ movements, abnormal growths within organisms, such as tumors and metastatic lesions, arthritic joints, and disorders in the cardiovascular system.

This section will deal with the impact that microstructured electronics in the 1980s will make on the development of (1) a new advanced generation of CAT scanners, and (2) the radionuclide ventriculogram system (RNVS), which is a new technique that provides a noninvasive substitute method for the cardiac catheterization angiographic procedure to diagnose cardiovascular diseases and cardiac disorder.

A. The CAT Scanner

The CAT scanner consists basically of an x-ray image system taking sequential exposures of a cross section of the anatomy. Computerized units integrate, perform pattern recognition analytical functions, and provide signals to reconstruct images that are presented on videotape and an associated data printout.

Recently, distribution of CAT scanners has decreased due to their high cost and the restrictions by state health control organizations which must approve health care facilities expansion, including increased bed capacity or instrument systems exceeding $100,000. The present market price for the CAT scanner is on the order of $350,000. Furthermore, at present, the scanners are cumbersome and must be housed in the radiological section of a hospital requiring a minimum area of 500 ft.^{2}

The first generation of CAT scanners took 6–8 min to generate an image with conventional integrated circuit components in the computer. The second generation of CAT scanners now employs 8-bit word and array microprocessors. Not only the size and the cost of this generation of scanners, but also the time of image production from 6 to 10 min to 5 to 20 sec have been reduced. Each cross-sectional view of the image produced by the scanner has $(256)^2$ to $(512)^2$ picture elements (pixels), which are reconstructed from several hundred different viewing angles. Furthermore, the present CAT scanners generate two-dimensional images and are too slow to identify the motion of organs such as the heart or the lungs in sequentially framed positions. In the present system, the patient must hold his breath to develop cross-sectional images of the lungs. To develop cross-sectional images of a beating heart is not within the capability of the present CAT scanners.

It will require a new generation of reconstruction processes, more efficient than the general-purpose array processors presently available, to accomplish extremely fast, real-time image construction.

The federal government is now supporting the next generation of x-ray CAT scanners at the Mayo Clinic [12]. The system under development is shown in block diagram form in Fig. 2. Twenty-eight rotating x-ray sources are arranged around a semicircle radius of 143 cm, independently controlled. Inside the circle is a 30-cm-wide fluorescent screen in a 58-cm radius. The signal flow paths are clearly shown in the diagram. Reconstruction of each cross section of the image is produced by the 28 image intensifiers and x-ray sources, multiplied and stored on video disks. Reconstruction of each cross section is replayed from the disks, digitized, and then fed into the general-purpose computer by a hard-wired digital construction unit, generating three-dimensional level displays. The aim of the system is to provide high control resolution, cylindrical scanning, and multiaxial tomographic capabilities that can scan up to 240 1-mm-thick

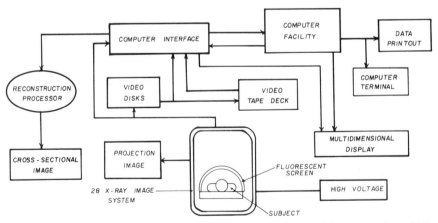

Fig. 2. System organization of advanced computerized axial tomography (CAT scanner).

cross sections in 11 msec. The system must repeat the complete scan procedure in intervals of 1/60th of a second and reconstruct the entire three-dimensional volume of the whole organ as well as dynamic changes in shape and dimensions of moving structures. No currently available general-purpose computer, operating unaided, can fulfill this system's data processing requirements at reasonable cost. Special-purpose arithmetic processors specifically optimized to execute appropriate reconstruction algorithms must be developed.

Such capabilities will only be possible with the new generation of VLSI microprocessor components that will be available in the 1980s, that will have speed capabilities well beyond 1,000 megaflops, ROM and RAM capabilities at least of 1 megabyte with 64-bit word sizes, with the possibility of 2 to 3 billion arithmetic operations per second and imaging processing times of about 1 nsec/pixel. This will produce computer-processing capabilities of the scan in a few seconds. The advanced microstructure solid-state components will be incorporated into the blocks in Fig. 2 designated as the reconstruction processor, the computer interface, and the computer facility.

The system, when effectively developed, will serve as a national resource so that biomedical researchers and clinical investigators will be able to study the advantages of real-time, volumetric scanning of entire organs and their blood flow patterns and compare the results of machines that produce but a single cross section over the duration of many seconds. Having proved its viability and importance in advancing medical diagnostic procedures, then it is reasonable to expect that on a basis of large-scale production and continued microstructured electronics engineering, a system will be developed that will certainly cost less than the million-dollar range for the pilot system described and be more generally available to hospitals and medical centers for highly sophisticated diagnoses of the diseases that today remain a mystery.

B. Computerized Nuclear Medical Systems

Not only will the next generation of CAT scanners revolutionize diagnostic processes, but other nuclear imaging techniques will provide non-invasive methods to study physiological processes, as well as be able to diagnose and understand the etiology of diseases. VLSI devices will offer a generation of computers that will make available, at low cost, miniaturized systems to the nuclear medicine specialist, particularly for the diagnosis of cardiac disease and dysfunctions. One such system is known as the Radionuclide Ventriculogram System (RNVS) [13]. This system will replace the cardiac catheterization angiographic procedure presently routinely applied to diagnose cardiac conditions. This latter procedure involves the injection of a dye and invasion of the arterial system with a catheter that is snaked through one of the vessels in the arm and into the heart. The procedure is risky, slow, and traumatic to the patient. The RNVS, however, devises a noninvasive technique to diagnose cardiovascular disorders without resorting to the in vivo cardiac catheterization method.

Both the presence of coronary artery disease and cardiac dysfunction can be detected with RNVS. Presently, in order to perform this test, the patient has to be brought into the radiological area, where use is made of equipment and instrumentation that is expensive and somewhat cumbersome. In this system blood is labeled by injecting $^{99}Tc^m$. A scintillation camera functions as a detector by gating the patient's ECG (electrocardiogram), and forms an image. The patient's R wave, used as a trigger, is sensed by the computer during each cardiac cycle, resulting in data with program time periods of 25 msec. Approximately 1000 cardiac cycles are analyzed and the processed integrated data represents both the spatial and temporal changes in blood volume during one heartbeat. These changes are viewed on two gated radionuclei angiocardiograms, one showing an image during diastole (relaxation of the heart) and the other during systole (the contraction of the heart). Further processing of the data results in a cineangiogram which places in evidence cardiac motion and global and regional measurements of cardiac functions such as ejection fraction, ejection rate, and ejection time.

A third system in this family of computerized nuclear diagnoses uses the transaxial computer tomography (TCT) [14]. This system provides a noninvasive quantitative measure of glucose metabolism, regional blood flow, and oxygen utilization. It enables one to develop a three-dimensional construction of the biodistribution of a radio tracer injected into the blood stream.

These two latter systems are still in their infancy but will play a major role in the delivery of health care and research. Presently, the cost of these instruments is high and the space requirements are significant. Miniaturization is the solution for these constraints. VLSI technology has already begun to be explored in advanced generations of these two systems, which not only will reduce the cost, but will also provide sufficient reduction in size to make portable these radionuclei computerized diagnostic systems so that they may be brought to the patient's bedside or any of the critical care areas of the hospital, such as the operating room, coronary care unit, or the emergency room.

The three systems described in this section, because of advances made in microstructured technology, will make available, at reasonable cost and size to medical practitioners and researchers, powerful tools that will enable them to venture into the deep realms of the cellular and structural dynamic organ systems and to explore the etiology of diseases that today are still a mystery. For example, the basic cause of coronary artery disease is unknown. Only risk factors for this disease are speculatively identified, and these factors change from year to year as do the standings in a popularity poll. The etiology of cancer is one of the deepest enigmas facing medical science. It is possible, through advanced, sophisticated instrumentation described in this section, that these mysteries may be solved.

IV. PROSTHETIC AND ORTHOTIC DEVICES

With the promise of inexpensive VLSI microcomputers, cost-effective production, and low-cost programmable software systems, the greater needs of individuals suffering from a wide variety of disabilities will be served. In particular, microstructure electronics will play a significant role in the development of closed-loop, electronically controlled, artificial limbs.

A. Artificial Limbs

In above-the-elbow prostheses, control signals are obtained from electromyographic muscle potentials (EMG) of various residual muscle groups which ordinarily activate arm movements. These signals are fed into an electronic system where they are amplified and processed. The signals then activate motors that effect motion within the prosthesis. Prior to the development of integrated circuits, the electronics for such artificial limbs had to be placed in a package external to the prosthesis, usually worn on a belt arrangement. Presently, in the classical integrated circuits used in the control of the sophisticated, available, upper-limb prosthesis, called "the Boston Arm" [15], limitations exist in processing the EMG signals and the flexibility of control. These circuits now are housed within the prosthesis proper, which sets limits on size and energy. Furthermore, these systems have to be calibrated from external devices.

With the advent of LSI and VLSI, greater versatility and refinement of control of upper-arm and other limb prostheses will be possible through improved selectivity of the control signals and selective integration of the information contained in the signals. At the same time, decreased size of the electronics and reduced power requirements will occur. An advance system of this kind is described in block diagram form in Fig. 3. In this system, instead of using an array of several electrodes as is necessary in

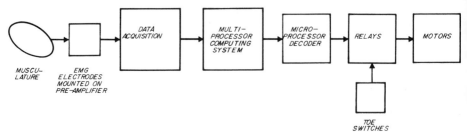

Fig. 3. System organization of above-the-elbow, computer-controlled, VLSI multiprocessor artificial arm.

present systems, only one or two electrodes will be required. The EMG electrodes can be directly mounted on a preamplifier attached to the musculature of the residual limb to provide control and discrimination of signals. The amplifier signals are then filtered in the data acquisition unit. The microcomputing system processes signals with integrating, separating, and pattern recognition techniques. The computer signal is interfaced with the prosthesis through a decoder unit that latches the binary function code into an output port, which, in turn, controls relays that switch power to the motors. The toe switches provide for the gross on–off control of the prosthesis.

B. Speech Synthesizers

VSLI will permit the development of analog microprocessors to combine analog/digital conversion-performing signal processing such as digital filtering and fast Fourier transformers with analog outputs on one chip. This is important for speech processors and devices into which one can talk and respond and will enable low-cost devices, simple to operate, to be developed for blind people to provide voice data entry and output [16].

Another system under development in this category is the hands-free artificial larynx. This is a device that can substitute larynx tones that are generated by a miniature sound source inside of the oral cavity. The sound is then sensed by a miniature microphone within the mouth. Audible speech is produced by a concealed flat unit consisting of a subminiaturized amplifier and loud speaker strapped to the user's chest. Further voice synthesizers applying an electronic keyboard and video displays will enable nonspeaking operators to see and correct messages, which are then translated into instructions for a voice synthesizer that speaks the message.

It will be possible to develop auditory prosthetic systems for the sensory deaf. This will employ implantable devices, mounted subcutaneously in the mastoid cavity, with data and power fed to it transcutaneously. The device will be able to translate sounds into a suitable pattern of electrical stimuli, which is sent to the central nervous system by way of the eighth nerve. This could make speech intelligible to one who is sensory deaf.

C. Cardiac Pacemakers

The cardiac pacemaker is a well-known device for the treatment of diseases causing rhythm disturbances in the cyclic functioning of the heart. It consists of a packet, implanted subcutaneously, that contains a power source in the form of a lithium oxide battery and an electronic unit. The electronics generates repetitive pulses to excite the electrical conduction pathways to the heart [17]. In demand-type pacemakers, the electronic unit contains a second channel whose function is to sense a normal heartbeat from the R wave of the electrocardiogram to detect natural contractions and, thus, inactivate the stimulating circuit. Leads are attached to the packet, which are connected either transvenously or epicardially to the heart.

In the programmable pacemakers, a pickup coil or reed switch inside the pacer is designed to receive an external signal, which can change its pulse rate, pulse amplitude, and sensitivity. The latest models of programmable pacemakers are capable of eight different pulse frequencies and eight different amplitudes and duration characteristics, in addition to twelve sensitivity settings. In order to change these characteristics, an external signal modulated either in digital form on a radio carrier or from a magnetic device, must be externally applied. Furthermore, with the presently available one-ventricle control pacemaker, only 80% of the contraction capability of the heart is utilized.

The future objective of pacemaker technology is to develop a unit that will automatically adjust its frequency, pulse amplitude, and duration in accordance with the physiological needs of the cardiovascular system. Instead of being able to pace one ventricle of the heart as is the case with present-day devices, it would be more desirable to pace the atria and ventricles sequentially. This would considerably increase blood flow from the cardiac chambers. With VLSI devices on the horizon, it will become feasible to develop the electronic circuitry to accomplish these desired characteristics and still maintain high reliability and small size specifications without increasing costs.

V. THE HEMODIALYZER: THE ARTIFICIAL KIDNEY

Until the advent of the artificial kidney in the 1960s, tens of thousands with diseased kidneys hopelessly died from uremic poisoning. Since then, treatment for the failing kidney through hemodialysis and hemofiltration has prolonged the lives of hundreds of thousands. The first dialyzer was constructed from a clothes-washing machine connected to the patient by means of invading an artery and vein with a large syringe needle whenever treatment was necessary. The system was highly unreliable and traumatic to the patient because of the repeated invasion of the circulatory system and loss of blood [18].

In the hemodialysis process, blood from an artery flows next to a membrane. Undesirable elements in the blood, e.g., urea, creatinine, uric acid, and phosphates, diffuse across the membrane into a saline solution (dialysate), which continuously flows past the reverse side of the membranal structure. The mass transfer of a chemical species, e.g., urea, depends on the blood and dialysate flow rates, the membrane characteristics, and the design of the dialyzer. Because the blood pressure is maintained higher than the dialysate pressure, water is ultrafiltered from the blood [19].

When, in the early 1960s, permanent surgically implantable arterial–venous (AV) shunts and fistulas were developed, hemodialysis became an accepted clinical treatment for final-state renal failure [20,21]. Then the system consisted of a unit that connected to the permanently implanted AV shunt with tubes. A roller pump propelled the blood through the unit and back to the venous system. The units had a layered membrane geometry to provide the interface between the blood and the dialysate for the diffusion of the undesirable elements from the blood. Early artificial kidneys had the dialysate fed from an individual tank. In the early seventies, methods to supply dialysate from a central station were developed. Not only was the apparatus crude from the standpoint of flow control, diffusion, and ultrafiltration processes, but available dialysis units plus services and facilities hardly met the needs of the number of patients who critically needed treatment. Thus a situation developed where patients were selected for treatment by a panel consisting of an attorney, a clergyman, and a physician, based on sociological, economic, and age factors and their contribution to society. Actually, this panel held the power of life or death over patients who needed dialysis.

Beginning in the late 1960s and early 1970s, hemodialysis units were produced in greater quantities. Electronic controls were introduced for flow and filtration rates. The federal government, through NIH, began to support research toward improvement of artificial kidneys. A new membrane geometry was developed for the diffusion and interchange of the undesirable components in the blood using disposable hollow fiber tube assemblies [22]. The government started to support treatment through the Medicare program. As a result of these measures, types and production of dialyzers proliferated and treatment centers sprang up in hospitals and private clinics. There no longer existed a shortage of units in clinics to provide treatment for kidney patients.

During the 1970s, with the advent of integrated circuitry and later microprocessors, sophisticated computerized controls and regulators of dialyzer functions were introduced into the units. These devices are open-loop controllers, monitoring blood flow rate, dialysate flow rate, transmembrane pressure, and ultrafiltered volume, which were set for an individual dialysis treatment. With the advent of microstructure electronics (VLSI), it is now possible to develop a computerized system that has incorporated within the dialyzer unit, unique closed-loop controls of the various parameters involved in the treatment, such as dialysis control, dialysis setting, ultrafiltration rate, temperature, venous blood pressure levels, arterial blood pressure levels, and blood conductivity. These parameters are indicated within the dialyzer units shown in Fig. 4. The chemical sensors and dynamic transducers for these parameters are incorporated into the unit with the levels controlled from a remote programmer.

VLSI can provide the advanced and sophisticated circuitry needed to allow the development of a system that incorporates individual treatment units in dialysis clinics with hospital computerized information systems. A programmed dialyzer that interconnects in an overall system is also shown in block diagram form in Fig. 4. The dialysis unit will have the

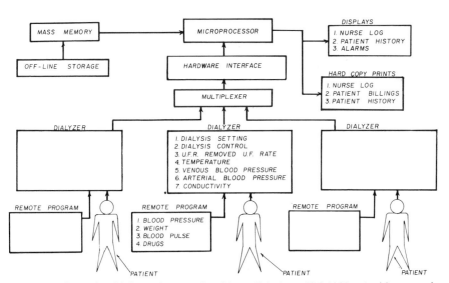

Fig. 4. Control and information transfer of hemodialysis (artificial kidney) with a central micro-main-frame computer.

capability of storing treatment parameters in the hospital's main computer and automatically log them with a hard copy printout. Built-in alarms could be incorporated, to be activated in both a central station and in the treatment center. This advanced composite system will improve the cost-effectiveness of the dialysis treatment, because the parameters, particularly those of quiescent levels of urea, bicarbonate, and ultrafiltration rates, will be carefully evaluated. Thus the termination of treatment can be uniquely identified. Therefore, the time of treatment will be reduced and treatments automatically individualized to the patient's needs, resulting in decreased costs and making the system available to a greater number of patients per unit time. The system in Fig. 4 is not yet in existence but is under consideration by industry, awaiting the availability of VSLI devices and microprocessor components.

VI. COMPUTER-BASED BIOFEEDBACK SYSTEMS

With the advent of VSLI technology, it will be feasible to develop a highly versatile, low-cost, portable, field-programmable computerized system with a biofeedback capability to allow medical practitioners to optimize a regimen of treatment therapy to improve the conditions of patients with neuromuscular deficiencies, stress-related symptoms, insomnia, and possibly organic diseases.

In biofeedback, sensory analog information is obtained from electrodes and transducers attached to a patient. These analog signals are amplified and fed into an analog/digital converter and then a microprocessor. Output information is presented to the patient, in the form of a video presentation, an audio signal, and/or an array of flashing lights. The patient, being in the loop, can cause these output signals to be continuously modified to reach a desired predetermined programmed level [23,24].

Presently, the most advanced computer-based biofeedback systems consist of several diverse components, which include discrete 8-bit microprocessors, A/D converters, audio signal generators, isolation amplifiers, memory, and digital interface circuitry. Typically, such systems occupy several logic boards and are physically bulky. These systems are only suitable for laboratories and clinical settings because they lack portability.

The sensory biofeedback systems of the 1980s will be able to capitalize on the integration of system functions in a microstructured array made possible by VLSI, as shown in Fig. 5, where it is seen that on one monolithic block several substrates can be joined by microcircuit interconnec-

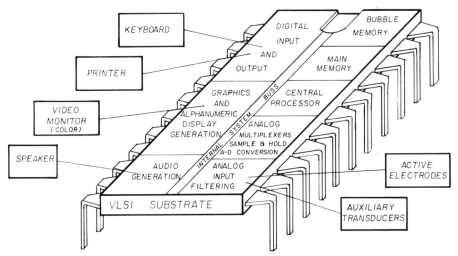

Fig. 5.   Electronics of biofeedback system on one monolithic substrate.

tions, each performing the functions of the units previously identified as discrete modules. These biofeedback systems will combine the data acquisition, signal processing, and data display functions on a single monolithic substrate instead of the relatively cumbersome systems produced from separate components. The resulting low-cost, miniaturized system will not only be better suited for clinical environments but will provide systems that are conveniently portable and appropriate for home use.

Furthermore, with the greater versatility and flexibility provided by the added electronic capacity offered by VLSI devices, different forms of proprioceptive and psychological feedback can be combined and correlated to provide new external and beneficial feedback pathways. The use of sensory pathways will proliferate as smaller and more inexpensive units become available to encompass such diverse areas as treatment of stroke victims, psychopathic disorders, epilepsy, and the control of organic diseases. The key factors to keep in mind in the application of VSLI are that the sensory feedback systems will become low in cost and be readily available to a larger market. This is additionally important inasmuch as sensory feedback therapy is now practically underwritten by federal health care agencies.

REFERENCES

1.   D. A. Simmons, "Medical and Hospital Control Systems: Critical Difference." Little, Brown, Boston, Massachusetts, 1972.

2. D. Bryson *et al.*, Megabit bubble memory chip gets support from LSI family, *Electronics* **52(9)**, 105 (1979).
3. A. Budkin, J. D. Dailey, D. Hunt, L. E. Bechtel, and B. F. Harvey, The Miami Heart Institute information system, *J. Med. Syst.* **2(1)**, 45 (1978).
4. W. J. Tompkins and J. G. Webster, "Design of Microcomputer-Based Medical Instrumentation." Prentice-Hall, Englewood Cliffs, New Jersey, 1980.
5. C. A. Titus, The impact of microcomputers on automated instrumentation in medicine (advances in architecture and software), *Ann. Symp. Comput. Appl. Med. Care, 1st, Washington, D.C.* (1977).
6. M. Bertrand and J. C. Foiret, Microprocessor application for numerical ECG encoding and transmission, *Proc. IEEE* **65(5)**, 714 (1977).
7. J. Zimmerman, MUMPS applications and their transfer, *in* Clinical Medicine and the Computer (*Proc. Ann. Conf. Soc. Comput. Med., 4th*) (M. A. Jenkins, ed.), Sect. 2, p. 2.7. Northern Graphics, Minneapolis, Minnesota, 1974.
8. E. R. Gabriel, Medical information systems, health records, and knowledge banks, *Med. Instrum.* **12(4)**, 245 (1978).
9. T. J. Lincoln, Computers in the clinical laboratory: What we have learned, *Med. Instrum.* **12**, 233 (1978).
10. R. T. Capece (ed.), Electronics—Looking ahead to the year 2000. *Electronics* **53(9)**, 448 (1980).
11. R. A. Brooks and G. D. Chiro, Principles of computer assisted tomography (CAT) in radiographic and radioisotopic imaging, *Phys. Med. Biol.* **21**, 689 (1976).
12. R. K. Jurden *et al.*, Electronics in medicine, *IEEE Spectrum* **16**, 76 (1979).
13. B. L. Holman, Nuclear imaging—medicine and technology at the crossroads, *Med. Instrum.* **13(3)**, 144 (1979).
14. M. E. Phelps, E. J. Hoffman, N. A. Mullani, C. S. Higgins, and M. M. Ter-Pogossian, Application of annihilation coincidence detection to transaxial reconstruction tomography, *J. Nucl. Med.* **16**, 210 (1975).
15. R. W. Mann and S. D. Reimers, Kinesthetic sensing for the EMG controlled "Boston Arm." *Man Machine Syst.* **11(1)**, 110 (1970).
16. G. D. Smith and H. A. Mauch, Research and development in the field of reading machines for the blind, *Bull. Prosthet. Res.* **10–27**, 62 (1977).
17. G. D. Green, "The Assessment and Performance of Implanted Cardiac Pacemakers." Butterworths, London, 1975.
18. W. J. Kolff and H. T. J. Berk, The artificial kidney: A dialyzer with great area. *Acta Med. Scand.* **117**, 121 (1944).
19. R. A. Gotch, J. Autian, C. K. Colton, H. E. Ginn, B. J. Lipps, and E. Lowrie, The Evaluation of Hemodialyzers. U.S. Dept. of Health, Education, and Welfare, Publ. No. (NIH) 72-103 (1972).
20. W. Quinton, D. Dillard, and B. H. Scribner, Cannulation of blood vessels for prolonged hemodialysis, *Trans. Am. Soc. Artif. Intern. Organs* **6**, 104 (1960).
21. J. E. Cimino and M. J. Brescia, Simple venipuncture for hemodialysis, *New England J. Med.* **267**, 608 (1962).
22. F. Gotch *et al.*, Chronic hemodialysis with the hollow fiber artificial kidney, *Trans. Am. Soc. Artif. Intern. Organs* **15**, 87 (1969).
23. J. Stoyoa *et al.* (ed.), "Biofeedback and Self Regulation," Vol. 1(1), pp. 1–145. Plenum Press, New York, 1976.
24. M. James, Micro-computers in biofeedback, *Ann. Symp. Comput. Appl. Med. Care, 2nd, Washington, D.C.* (1978).

Chapter **7**

Signal Processing Using MOS–VLSI Technology

R. W. BRODERSEN

Department of Electrical Engineering
and Computer Sciences
University of California
Berkeley, California

I. INTRODUCTION

Very large scale integrated circuits (VLSI) is the name given to integrated circuits (ICs) that, by advances in IC technology, have become

189

larger than some arbitrarily chosen size. The point at which circuits become VLSI instead of just LSI is unclear. This confusion is apparent if one observes the wide range of definitions for VLSI that have been proposed, which range from as low as 1000 logic gates on one integrated circuit (chip [1] up to circuits containing more than 100,000 gates [2]. The reason for this indecision is that VLSI is not a step function improvement in the capability of IC technology but rather just the next stage in its continuing development. This continuity makes it possible to use the experience of the recent past to project the nature and timetable of future developments.

A critical assumption must be made about what present-day technology to use for the projections. Most (but by no means all) believe that the metal–oxide–semiconductor (MOS) technology has the most promise for the future because of the relatively simple structure of the MOS device. The more complex bipolar technology and devices are not as able to exploit the advantages in density of shrinking (scaling) the minimum feature size. However, in order to actually provide new capabilities for signal processing, it is speed, not density, that is required. Increased density is important in that it provides the possibility of reducing the size, cost, power, and weight of systems which are now implemented with bipolar technology; but even in 10 years, when MOS technology is near its fundamental limits, its processing speed will not be as high as today's advanced bipolar technologies. As will be dicussed in this chapter, this lack of capability even occurs if it is assumed that the architectures and algorithms (such as parallel processing and pipelining) that are used can convert increased complexity into increased processing power.

The reason high speed is so important is that signal processing is the computation of mathematical algorithms on "real-world" signals in real time; i.e., the processing rate is equal to or greater than the data rate of the incoming signal. The real-time requirement has a very fundamental effect on the hardware implementation since processing speed, even for the lowest bandwidth signals, is usually the most difficult performance criteria to meet. Until recently the only means for obtaining adequate speed in cost-sensitive applications has been the use of analog circuitry, even though these techniques severely limit the complexity of the signal processing algorithms that can be implemented. It is expected that the capabilities of VLSI technology will finally make it possible for digital techniques to be useful in other than low-volume, expensive systems which can make use of high-speed bipolar technology.

However, it should be mentioned that the imminent impact of digital processing has been predicted by researchers for at least the last 10 years. Except in very limited application areas, the analog approach has been found to yield adequate performance at a much lower cost, less power, and smaller size. Recent advances in analog techniques have been made which use a discrete-time representation of signals and can be implemented using MOS technology [3]. Thus these analog methods can make use of many of the advances in digital signal processing theory as well as the advances in IC technology. In fact, to a very important extent, progress in digital signal processing is tied to advances in analog techniques, since the "real-world" signals must be amplified, filtered, and A/D converted in the analog domain before they can be processed digitally.

For signal processing, probably the most important system consideration is the bandwidth of the real-world signals to be processed. A graph of the bandwidths of a variety of signals is given in Fig. 1. The bandwidths in this figure cover the enormous range of 10 orders of magnitude in frequency. At the low end are the seismic signals which do not extend much below 1 Hz because of the absorption characteristics of the earth. At the other extreme are the microwave signals which are not used much above 30 GHz because of the difficulties in performing even the simplest forms of signal processing at higher frequencies.

In order to perform signal processing over this range of frequencies, a variety of techniques have been developed, which are almost exclusively analog above 10 MHz and digital below 100 Hz, as shown in Fig. 2. In the overlap region a trade-off must be made between the accuracy and flexibility of a digital approach and the low cost, power, and size of analog techniques.

The fastest-moving boundary in Fig. 2 is the upper limit of the MOS–LSI digital logic, which, as the technology progresses to VLSI (as

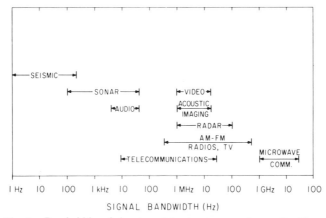

Fig. 1. Bandwidths of signal used in signal processing applications.

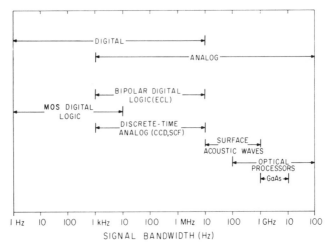

Fig. 2.   The signal bandwidths that can be processed by present-day technologies.

will be shown in a later section), should be able to process signals by the year 1990 at bandwidths of 1 to 10 MHz. This will be accomplished by means of the utilization of greatly increased density along with increased device speeds. However, as can be seen in this figure, bipolar digital techniques are already able to process at these rates. Therefore, VLSI will not make possible increased processing rates over what can now be achieved with processors built with bipolar technology, but rather the reduced cost, size, and power requirements of the MOS–VLSI signal processors will be the primary advantage. This will make it possible for signal processing techniques to make an impact in cost-sensitive areas such as consumer products, which has, until now, been unable to afford the costs of using sophisticated signal processing techniques.

II. SURVEY OF SIGNAL PROCESSING ICs

At present there are available a variety of ICs which were primarily designed for signal processing. They make use of MOS technology to implement a significant portion of a system function on a single chip. An investigation of the characteristics of these chips is probably the best guide of what to expect in the near future. They can be broken up into three basic groups; general-purpose digital processors, special-purpose digital processors, and special-purpose analog processors.

A. General-Purpose Digital Processors

The general-purpose digital processors are, in most cases, relatively straightforward extensions of conventional general-purpose microprocessor architectures with the addition of a multiplier, as shown by the architecture of a recently developed chip in Fig. 3. At this time there are four processors that have been announced and some of their more important characteristics are presented in Table I. Processors 1 and 2 were first announced several years ago, while processors 3 and 4 were announced this year and, therefore, their increased capability is partially due to the use of more advanced technology.

Processor 1 is the only one of the four that performs multiplications by software control [4]. This places an extra undesirable burden on the programming of this chip compared to the others. However, it contains a 9-bit A/D and D/A converter which is an attempt to address the interfacing problem between real-world signals and the digital processing. Unfortunately, 9 bits yields a dynamic range of only 54 dB, which is inadequate for most applications.

Processor 2 contains an integrated array multiplier (i.e., an array of logic gates that performs a multiplication with one instruction), but the maximum word size is only 12 bits at the multiplier inputs [5]. For most signal processing algorithms this is insufficient accuracy, since at internal steps in many algorithms much higher accuracies are required (typically 16 bits is a bare minimum).

TABLE I

Comparison of Major General-Purpose Digital Processors

Characteristics	Processor 1	Processor 2	Processor 3	Processor 4
Device number	Intel 2920[4]	AMI 2811 [5]	Bell Laboratories [6]	Nippon Electric Co. [7]
Chip size (mm²)	29	25	68	28
Number of transistors	~30,000	30,500	45,000	40,000
RAM (bits)	1K	2K	2.5K	2K
ROM (bits)	4.5K	6K	16K	18K
Clock rate (MHz)	~2.5	~3	5	8
Size of multiplier	Software	$12 \times 12 \Rightarrow 16$	$16 \times 20 \Rightarrow 40$	$16 \times 16 \Rightarrow 31$
Power dissipation (W)	0.8	1.0	1.5	0.9
Process technology	NMOS $L = 4.5\ \mu m$	VMOS	NMOS $L = 4.5\ \mu m$	NMOS $L = 2.5\ \mu m$

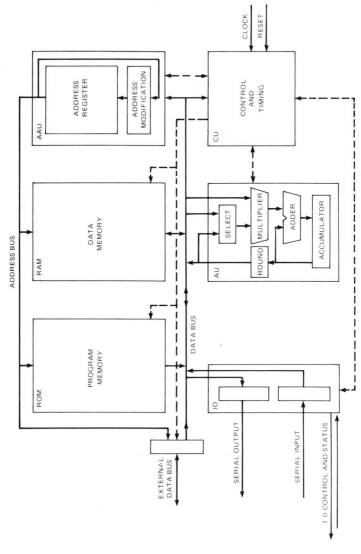

Fig. 3. Architecture of digital signal processor [6], developed for telecommunications applications. This is processor 3 in Table I.

the telephone industry [6,7]. The interfacing problem is thus solved since these chips can use the very inexpensive A/D and D/A converters that use this code which are produced by a variety of manufacturers. Both of these chips have adequate internal precision and speed to perform sophisticated signal processing at telephone bandwidths (4 kHz). The major difference between these two chips is that processor 4 is fabricated using a technology that has a minimum linear dimension which is 60% of that which can be used in processor 3. This shrinking (scaling), which will be discussed in the next section, results in a chip that is smaller, faster, and less power-consumptive.

B. Special-Purpose Digital Processors

The greatest impact that VLSI will probably have on signal processing is through special-purpose digital chips. When a chip is designed for a specific application, it is much easier to optimize the architecture in order to fully utilize the increased number of devices that VLSI technology will provide. Also, by completely integrating entire system functions, the advantages of cost, size, power, and weight can be fully realized since general-purpose chips usually will require a number of external components. Two examples of chips of this type are a speech synthesizer, which was introduced in 1978 by Texas Instruments [8], and an echo canceler, which was recently developed by Bell laboratories [9].

The speech synthesizer chip uses a sophisticated speech coding technique based on linear prediction (LPC). This method makes it possible to reduce the bandwidth of speech so that high-quality speech can be generated from data rates of less than 2400 bits/sec. Even though the theory and hardware (high-speed bipolar processors) of adequate capability to implement the algorithm have been in existence for about 10 yr [10], the application of LPC has been very limited. This changed essentially overnight, however, when the entire synthesizer was integrated onto a single chip using MOS technology and it was possible to generate over 100 sec of speech at a cost low enough to be sold as a toy. The technology actually used in this speech synthesizer was 4–6 yr behind the state of the art, even when it was introduced in 1978. Adequate performance was obtained by use of a pipeline architecture, which trades off device speed with additional circuit complexity.

The echo canceler chip implements a self-adaptive cancellation algorithm that supresses echoes on long-distance telephone calls. Before development of this chip, the per channel costs for this function were too high for wide-spread application. This chip was also fabricated using a

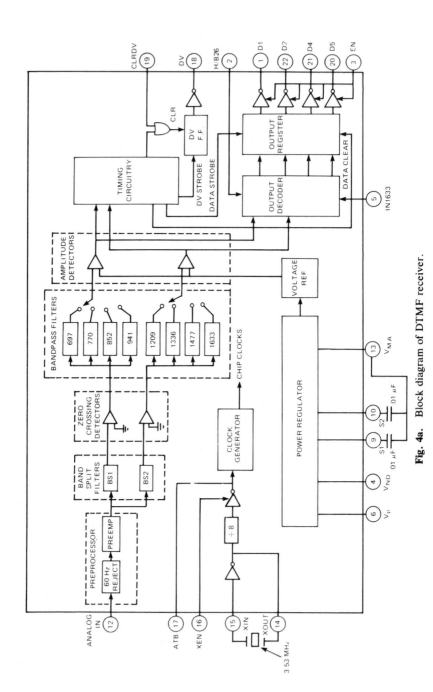

Fig. 4a. Block diagram of DTMF receiver.

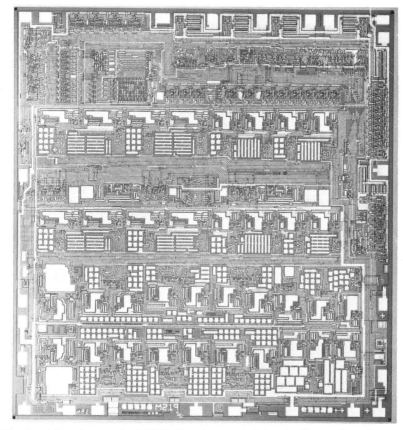

Fig. 4b. Microphotograph of MOS IC analog-sampled data implementation of receiver.

conservative technology but is near the limits in complexity at 35,000 transistors. A standard cell layout technique was used (see Section IV.E.2), which made possible very short development time (<6 months) and a design that was functional on the first pass.

C. Special-Purpose Analog Signal Processors

To demonstrate some of the advantages of analog signal processing, several analog ICs will be described and compared to the capabilities of all-digital approaches.

Discrete-time analog techniques have proven to be especially efficient in implementing frequency-selective filtering (see section V.A). A number of chips have recently been developed using these techniques, which implement approximately 20 poles of filtering for antialiasing signal reconstruction, 60-Hz rejection, and power amplification for the PCM transmission of telephone signals. One of the more recent chips developed by MOSTEK Corp. for this application consumes a total of 20 mW when the chip is active (0.2 mW in standby mode) [11]. The sample rate for the filters on this chip is 128 kHz. If the general-purpose digital processors just described were operated at a 128-kHz sample rate, they would only be able to implement 2–3 poles of filtering. Even with this limited capability, they require more than 50 times the power, as well as 5 to 10 times the chip area of the analog approach (while using the same technology).

Analog-sampled data signal processors are capable of a reasonable degree of complexity, as demonstrated by the Dual Tone Multiple Frequency (DTMF) receiver developed by Silicon Systems [12]. The function of this receiver is to detect pairs of tones on a telephone line in the presence of dialing signals, speech, and line noise. The block diagram of this chip is shown in Fig. 4a and a microphotograph of the chip is shown in Fig. 4b. The IC uses analog techniques to implement about 40 poles of filtering as well as precision zero-crossing and amplitude detection. This chip also contains a considerable amount of digital circuitry which is used for formatting and timing the digital output signals of the chip as well as the system.

III. SCALING

One of the most important reasons for the increasing dominance of MOS technology is that as the device dimensions are reduced (scaled), improvements are obtained in essentially all performance characteristics of the resultant integrated circuits. Therefore, in order to predict the capabilities of the technology for the future, an understanding of scaling theory is required.

The three terminals of a MOS transistor (MOSFET) are the source, drain, and gate, as shown in Fig. 5. The gate of a MOSFET, which is shown pictorially in Fig. 6, is typically a layer of deposited polycrystalline

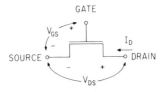

Fig. 5. MOSFET, showing voltage and current conventions.

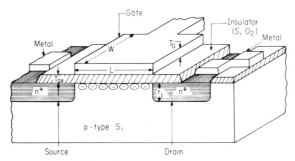

Fig. 6. Drawing of a silicon-gate MOSFET.

silicon (polysilicon), which is heavily diffused with impurities so that it is a reasonably good electrical conductor. It acts as one plate of a capacitor with the low-conductivity silicon substrate as the other plate. The source and drain are heavily doped regions that form $p-n$ junctions (diodes) in the silicon as well as forming high conductivity paths. Under all normal operation, the voltage on the source and drain regions with respect to the substrate is of the polarity that reverse biases the source and drain diodes. Under these conditions there is only extremely small current flow into the substrate. In order for there to be any current flow from the source to the drain, it is necessary to apply a voltage to the gate. This attracts carriers to the silicon–silicon dioxide interface and thereby forms a conducting channel between the source and drain regions. The gate of a MOS transistor, therefore, acts as a switch between the source and drain, which is open or closed depending on the voltage applied to the gate.

In Fig. 6 the physical dimensions are labeled, which, under standard scaling rules, are all reduced proportionally. The cross-hatched area is the insulator (SiO_2), which has a thickness T_{ox}; r_j is the junction depth of the source and drain diffusions (the shaded area); and the width and length of the gate are W and L, respectively. The metal leads in Fig. 5 (which are typically aluminum) can connect this transistor to others as well as to the package. If all of the physical dimensions are scaled down by a factor of K, then the density (number of circuits per unit area) of circuit elements can increase as K^2. This increase in density has, in recent years, been the greatest contributor to the continuing trend of increased complexity in integrated circuits.

For the past 20 yr it has been noted that the number of devices on a single integrated circuit has doubled every year [13]. This relationship is established by plotting the number of transistors per chip versus their year of introduction, as in Fig. 7. It was recently noted by Gordon Moore of Intel Corp., who first noted this trend, that in recent years there seems to be "slowing" to a doubling in complexity every year and a half, which

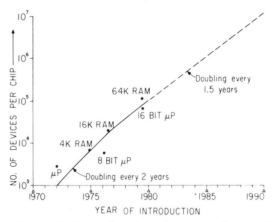

Fig. 7. Increase in the number of transistors on an integrated circuit.

is indicated by the dashed line in this figure [13]. From the extrapolation it can be seen that the complexity of integrated circuits is expected to increase by 100 times in the next 10 yr to chips that contain up to 10 million devices.

The simplest form of the current–voltage equations (see Fig. 5) of a MOSFET in which electrons carry the current in the channel (an N-channel or NMOS device) are given by

$$I_{ds} = \begin{cases} \mu_n C_{ox}(W/L)V'_{gs}V_{ds}; & V'_{gs} > 0, \quad V_{ds} > V'_{gs}, \quad (1a) \\ 0; & V'_{gs} < 0 \quad\quad\quad\quad\quad (1b) \end{cases}$$

where V_{ds} is the drain voltage, $V'_{gs} = V_{gs} - V_t$, μ_n the electron mobility, C_{ox} the oxide capacitance per unit area. The effective gate voltage V'_{gs} takes into account a series of charge layers beneath the gate which can be modeled as an offset voltage V_t, in series with the applied gate voltage V_{gs}.

An important parameter for signal processing is the delay due to the finite resistance of a MOSFET to charge the parasitic capacitances to which it is connected. A rough estimate of this time (typically referred to as the gate delay) can be found by calculating the resistance of the channel of a MOSFET using Eq. (1) to obtain

$$R = V_{ds}/I_{ds} = L/W\mu_n C_{ox}V'_{gs}, \quad (2)$$

and since the gate capacitance C_g that a transistor must drive is given by

$$C_g = C_{ox}LW, \quad (3)$$

the gate delay $T_d = RC_g$ can be found to be

$$T_d = L^2/\mu_n V'_{gs} \quad (4)$$

In the past few years since MOS circuits were made to operate from single 5-V supplies, the voltage V'_{gs} has been kept constant as the channel length has been scaled down by K (this is due to changes that will be discussed later). This results in a delay time inversely proportional to the square of the scaling factor K:

$$T_d^{scaled} = (L/K)^2/\mu_n V'_{gs} = T_d/K^2. \tag{5}$$

In Fig. 8 the minimum delay versus channel length L is plotted as determined by ring counters for a variety of scaled devices (the points labeled (a)–(g) correspond to references [14–20], respectively). A ring counter is a loop of an odd number of inverters N that oscillates at a frequency proportional to $2N$ times the delay of one inverter. This is a minimum delay value since the load of each inverter is only one other inverter. Also in this figure, for comparison to the ring counter data, are three points that correspond to access times of commercially available static random access memories. As seen in this figure, they also follow the speed increase predicted by Eq. (5), which is plotted as the solid line in Fig. 8.

In order to indicate the rate at which scaling is proceeding, in Fig. 9 the active channel length L is plotted for a variety of commercial processes as a function of the year of their introduction. If the past trend continues, then in 1990 the channel length should be approximately 0.4 μm, which approaches some fundamental limits of MOS technology [21].

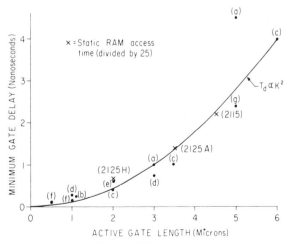

Fig. 8. Minimum gate delay and static RAM access times as a function of active channel length L.

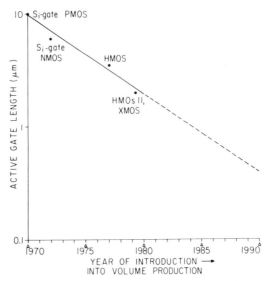

Fig. 9. Rate of decrease of the minimum active channel length L for commercial processes.

IV. IMPLICATIONS OF SCALING

In the last section it was shown that the device technology has been following a predictable trend of scaling to smaller dimensions, and it is expected that this trend will continue. This extension of MOS technology to VLSI complexity has some important implications for those who would like to use it for signal processing applications. In the following sections a few of these implications will be discussed.

A. New Voltage Standard

As mentioned previously, when MOS technology advanced to the point at which 5-V-only circuits could be used, it became almost a necessity for any new component to retain 5-V compatibility. In Fig. 10 a plot is given of the supply voltage versus channel length that was used for the delay times plotted in Fig. 8. There are several reasons for this inflexibility. The first is that even at the present levels of integration the power supply can be one of the most costly components of the system as well as being the most unreliable. Therefore, the use of the smallest number of the most standard power supplies is desirable from the viewpoint of the system designer. Since in the most complex systems there will be (and will continue to be) a relatively large amount of circuits using unscaled MOS circuits or other technologies which require 5 V, it is undesirable to use MOS parts that require additional voltage levels.

A major inconsistency between scaling the physical dimensions and retaining a constant supply voltage is that the electric fields across the insulator beneath the gate are increased. These increased fields result in electron injection and tunneling into and through the oxide as well as considerably enhancement of the phenomenon of oxide wear-out. All these effects can adversely impact the reliability of circuits. In Fig. 11 a plot of oxide thickness versus channel length for a number of recently developed processes is shown. If it were assumed that the oxide thickness continues to scale with channel length as projected by the line in this figure, then the oxide thickness for a 0.4-μm device would be on the order of 80 Å.This would result in oxide electric fields of 7×10^6 with a 5-V supply. In order to retain present-day reliability, the oxide electric field should be at least four times less than the breakdown electric field (which is approximately 10^7 V/cm for thin oxides). Thus the new supply voltage standard will need to be 2 V or less by the end of the decade in order for VLSI circuits to be widely used.

B. Processing Bandwidth

Extrapolation of the trend of scaling of the channel length plotted in Fig. 9 indicates that there should be a decrease by a factor of 5 by the year

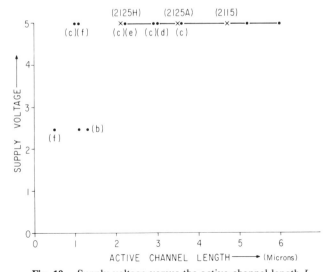

Fig. 10. Supply voltage versus the active channel length L.

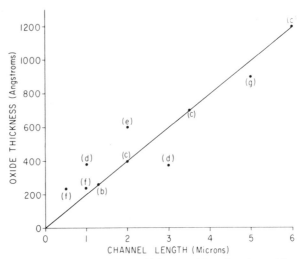

Fig. 11. Scaling of the insulator thickness with active channel length L.

1990. Since at 0.4-μm channel lengths the technology is approaching some fundamental limits, either fundamentally new technologies must appear or the progress that the IC industry has experienced will drastically be reduced. In either case it is not productive to attempt to project past 1990. In that year, from Eq. (5), we see that the gate delay should decrease by 5^2 or 25 times. However, this assumes a constant supply voltage, which, as previously discussed, will result in reliability problems. If it is assumed that the new supply voltage standard is reduced two and a half times to 2 V, then from Eq. (4) it can be seen that the speed will also decrease by that amount. The increase in bandwidth of signals that can be processed due to increases in device speed will, therefore, be a factor of 10.

The other approach to increasing bandwidth, as previously discussed, is to exploit the increased number of devices available on a VLSI circuit. From Fig. 7 it is seen that 100 times more devices will be available in 1990. Because of the difficulty in efficiently utilizing increased complexity, the processing rate will rise at a slower rate than the increase in the number of devices. One estimate for this effect is that the processing power increases as the 0.7 power of the number of circuits [22]. This would imply that the increased density of VLSI would result in a factor of $100^{0.7}$ or 25 times more processing power.

The total improvement in processing bandwidth from now to 1990 should, therefore, be on the order of 25 × 10 or 250 times. If the processing bandwidth of MOS–LSI at this time is taken to be 10 kHz, the bandwidth in 1990 should, therefore, be in the range of 1–10 MHz. As seen in Fig. 2, this is just equivalent to the present capability of bipolar processors, so the real advantages of the VLSI circuits will be in the low cost, size, and power of the VLSI approach.

C. Algorithms and Architectures

As discussed in the previous section, the actual increase in processing rate for VLSI, even at the fundamental limits of MOS technology, will not be as fast as current bipolar processors, so there cannot be much allowance for inefficient architectures.

It is, therefore, important that architectures be developed that can efficiently use the increased number of devices that will be available. To do this architectures make use of concurrent processing in a variety of forms, such as parallel processing and pipelining, in which simultaneous processing occurs in either multiple or single data paths, respectively. In order to use these architectures, the signal processing algorithms should contain as little branching and decision-making steps as possible. Because of this it is not true that the best signal processing algorithms are those that contain the minimum number of multiplies or other arithmetic operations, since they may be trading off arithmetic computation for more complex control. A better criteria would be strongly weighted toward those algorithms that can be implemented by parallel processing or ones that have a minimum of branching so that pipelining can be used. For example, the extra data manipulation that the fast Fourier transform algorithm requires may even make that extremely powerful algorithm less desirable than a straightforward calculation of the discrete Fourier transform, especially if only a small number of coefficients are required.

D. Access to VLSI

While most performance parameters improve as the technology is scaled, there is one parameter that rapidly gets worse and it is the capital costs of starting up a new production line. As indicated in Fig. 12 the cost of a state of the art, high-volume production line (greater than 50,000 wafer starts/month) has been increasing almost as fast as the increase in complexity of the circuits being fabricated. This is having some profound effects on the nature of the semiconductor industry since only the largest manufacturers are able to afford this capital investment. This also has a direct effect on the availability of VLSI chips for signal processing applications, since, in general, these chips are only needed in small volumes (at least small compared to memory and microprocessor chips) and it is not economical for a large manufacturer to produce them. This means that

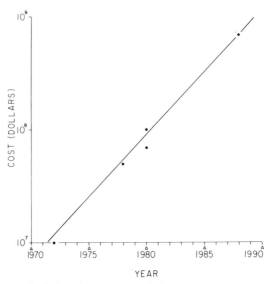

Fig. 12. Capital equipment costs of high-volume production lines.

system houses that do not have internal fabrication capability will find it very difficult to obtain VLSI chips designed for specific signal processing applications. Therefore, the vertically integrated system/IC manufacturer will be the primary developer of VLSI signal processing circuits. There will, of course, continue to be general-purpose chips that the IC manufacturers will continue to make available; however, systems developed based on these chips will have performance that is increasingly behind circuits that can directly use architectures to exploit the increased number of devices made available by VLSI.

Simultaneously with the pressures for system houses to develop in-house processing capability, there are also pressures forcing IC manufacturers to develop system expertise. A problem that has been voiced by a top-level manager of a components manufacturer is "besides memory I don't have the slightest idea how to take advantage of VLSI" [13]. In the search for new markets, some IC manufacturers are recognizing that signal processing applications have enormous potential. Recently there has been much interest in speech processing (recognition and synthesis) by the IC industry. Projections for this market segment are that total sales will reach the range of $1.3 to $1.8 billion by the late 1980s [23]. In order to attract personnel with systems backgrounds to the IC industry, some of the major manufacturers are organizing system-oriented subdivisions with a charter to develop speech processing products.

The other motivation for IC manufacturers to get into the systems business is that, because of the high cost of capitalization, the VLSI circuits will be very costly. Therefore, there is a strong economic advantage to making systems that can be sold at much higher prices (and profit margins) than individual components.

E. Design Methods

A critical problem related to the increase of device complexity is how a chip with an increase of two orders of magnitude in the number of devices can be designed in a reasonable time at a reasonable cost. Also, after it is designed, the layout must be checked to ensure that it is correct (verified). A standard method used at this time for verification is to trace the circuit by hand. This method is clearly inadequate to handle circuits containing 1–10 million devices. In fact, there are some who feel that the design problem is so difficult that only general-purpose circuits are viable, because their general-purpose character will allow them to be made in sufficient volume to justify the design costs. The problems with this approach are that it becomes increasingly difficult to define a chip that will be useful to large numbers of users and to define a general-purpose architecture that can exploit concurrent processing.

Fortunately, there are some intermediate methods between the past procedure of optimization of almost every device and the inefficiency of a general-purpose chip. Three of these techniques are gate arrays (masterslice), standard cell layout, and logic arrays. An important advantage to these design approaches is that one does not need to be very knowledgeable about IC technology in order to design a chip. This results, to some degree, in inefficiency in the use of the technology so that, for these design methods, the processing bandwidths in the year 1990 will be even lower than the 1–10 MHz previously projected.

1. Gate Arrays

The gate array or masterslice technique makes use of a chip which has a fixed array of very simple gates interconnected in a customized manner, using the various levels of interconnect available. Since the location of the logic gates are fixed, the only remaining step in the design is the interconnection of the available gates. High levels of computer assistance can be used in this wiring as well as verification.

Probably the most aggressive use of gate array techniques for signal processing applications is by Fujitsu Ltd., who have developed a family of MOS gate arrays containing up to 3900 gates (15–20,000 devices) in a technology that has a minimum gate delay of 8 nsec. By comparing Figs. 7

and 8, it is seen that these arrays are well below the capability of state of the art technology, but they compensate for this deficiency by the reduced design time and costs. It also should be pointed out that even the most advanced 16-bit microprocessors now available (e.g., Motorola 68000) cannot even come close to implementing the real-time signal processing functions (such as a 2400 baud modem) for which these gate arrays are being used.

2. Standard Cell Layout

In this technique a library of logic functions is developed which contains logic elements such as NORs, NANDs and flip-flops. These standard cell are typically rectangular and have standard locations for the input and output connections so that a high degree of computer assistance can be used for placement of the cells and routing of the interconnect lines. Standard cell layouts allow more flexibility than is available in the gate array approach yet retain sufficient structure that a high level of computer assistance can still be used. The use of standard cells is basically the extension to silicon of the techniques that have been in use for a number of years in design of small-scale integrated circuit logic in printed circuit boards.

Since the computer does the routing of the interconnections (often with considerable assistance by the designer), the verification problem is considerably reduced since a logic simulator can be used to check the logical operation of the circuit. Also, simulation of the circuit's speed performance can be done by high-level timing simulators since the device-level simulation has already been performed on the standard cells. The major disadvantage of this approach is that, due to the rigidity of the shape and placement of the building-block structures, there is a high premium paid for the area required for interconnections, which results in inefficient use of silicon area and long interconnection lines, which will yield increasing performance degradation (longer delay times) as the technology is scaled.

3. Logic Arrays

It is possible to directly implement random logic with read only memories (ROM). This becomes particularly attractive for VLSI since computer automation can be used in the design and layout. A particularly efficient use of ROMs is in a programmable logic array (PLA) configurations, in which the output of one ROM forms the input to another. There are powerful minimization programs which have been developed that reduce the size of the ROMs in PLA for a given logic function, which can result in PLAs that actually require less area than an optimized random logic layout.

ROMs and PLAs are now extensively used for control functions in large complex circuits. For example, more than one-third of the Motorola 68000, 16-bit microprocessor is ROM and PLA. Another recent investigation of PLAs was in their use for implementing large adders. It was found that the number of bits in the PLA increased as $(6.6N)^{2.5}$, where N is the number of bits in the adder [24]. This should be compared to the size of an adder implemented using conventional random logic techniques (ripple carry adder) that require $35N$ transistors [25]. Since each bit of a ROM can consume as much as 10 times less area than each transistor of a random logic layout, it is actually more efficient in area to implement up to a 13-bit adder using a PLA.

V. ANALOG SIGNAL PROCESSING

Since in the real world signals are analog and continuous in time, in order to use a digital signal processor there is a need for analog interface circuits, which perform filtering, amplification, analog-to-digital, and digital-to-analog conversion. In the past the interface circuits have often been an insignificant fraction of a complete system. However, with the projected densities and low cost of VLSI circuits, it is anticipated that for many system applications the digital portion of the system will be implemented on a single chip, which places new importance on the complexity of the interface circuitry that connects that chip to the outside world. Even now, with the density of digital circuitry achievable with state of the art LSI technology, the chip count, power dissipation and size of simple systems can be dominated by the peripheral interface circuitry [26]. For example, a typical data acquisition board for a commercially available microprocessor contains 10 chips and costs $100. Clearly, further reduction in the digital circuitry in the microprocessor, which can now be implemented in a single chip for less than $10, will have small impact on a system that makes use of such a data acquisition board. Therefore, to truly realize the advantage of even present-day technology, the importance of integration of interface circuits is clear. Extensive progress has been made in the development of interface circuits that use MOS–LSI technology and thus make possible the integration of the analog circuits onto the same chip as the digital processing. Also, because the density advantages of MOS carry over into analog circuits, sophisticated analog signal processing circuits can be implemented on a single IC. This makes it possible for the analog circuits to exploit the low cost, size, and power advantages of integration.

The primary emphasis in the development of VLSI technology has been on the optimization of the performance of digital logic circuits. This, unfortunately, is inconsistent with improving (or even maintaining) the performance of the analog interface circuits. It would be a considerable limitation to the usefulness of VLSI technology if techniques are not found tha make it possible for the interface circuits to improve in performance along with the digital technology.

A. Analog Components

Elements of key importance in MOS analog circuits are capacitors, resistors, switches, and amplifiers [3]. Ratios of capacitors are the precision component of A/D and D/A converters and also determine the frequency response of filters. The near-ideal MOS switches make possible novel circuit configurations not possible in other technologies.

1. MOS Capacitors

The MOS transistor itself is essentially a nonlinear capacitor in which the charge induced on one plate by the applied voltage forms the conducting channel of the transistor. The dielectric of this capacitor is silicon dioxide, which is one of the most stable dielectrics known. However, the MOS transistor itself is not particularly useful as a capacitor because of its inherently nonlinear behavior. Within an advanced MOS technology, capacitors having a sufficiently low voltage coefficient to be useful as precision passive components are generally made with two separate layers of polysilicon separated by SiO_2.

These capacitance structures tend to have common characteristics. From the standpoint of switched-capacitor filters and D/A converters, the most important of the characteristics of MOS capacitors are ratio accuracy, voltage coefficient, and parasitics.

a. Ratio Accuracy. A key aspect of the performance of any frequency-selective filter is the accuracy and reproducibility of the frequency response and, for a D/A converter, the accuracy of the quantized voltage levels. MOS-compatible circuit techniques have been developed which rely on a certain level of accuracy in the ratios of capacitors in order to meet these requirements.

Integrated MOS capacitors have a value that is determined by the dielectric constant, the thickness of the dielectric, and the area of the capacitor. Assuming that the dielectric constant and thickness do not vary, the ratio of two capacitors made within the same integrated circuit will depend only on their area ratio. This is primarily determined by the geometrical shape of the capacitors defined by the photolithographic mask used to make the integrated circuit. Errors in the geometrically defined ratio occur because the actual definition of the capacitor is done optically, which results in errors in the capacitor edge location due to the finite wavelength of the light used and the control of the chemical or plasma etching of the metallization and, in addition, because the thickness of the dielectric can vary with distance across the integrated circuit. These effects can be alleviated with careful layout of the components and the errors generally get smaller as the capacitor dimensions are made larger. Generally speaking, the achievable ratio accuracies range from 1 to 2% for small capacitor geometries (<400 μm^2) to on the order of 0.1% for capacitor geometries approaching the limit of economical size (40,000 μm^2) [27]. This also implies that as the capacitance ratio gets larger, the achievable accuracy in a given area decreases since the smaller of the two capacitors must be decreasing.

b. Voltage and Temperature Coefficients. The MOS capacitors made with heavily doped silicon plates display voltage coefficients in the range of 10 to 100 ppm/V. Temperature coefficients are generally in the range of from 20 to 50 ppm/°C and are, of course, much lower for the value of a ratio [27]. These variations are low enough to be insignificant in almost all applications.

c. Parasitic Capacitances. A sizable parasitic capacitance exists from the bottom plate of the capacitor to the substrate that is the capacitance of the SiO_2 layer under the first layer of polysilicon. Typically, this capacitance has a value of from one-fifth to one-twentieth of the MOS capacitor itself, depending on the specifics of the technology. Also, because the top plate of the MOS capacitor must be connected to other circuitry, a small capacitance will exist from the top plate to the substrate due to the interconnections. This capacitance can range from 0.01 to 0.001 of the desired MOS capacitance, depending on the capacitor size, layout, and technology. These parasitic capacitances are unavoidable, and the design of filters and A/D and D/A converters must be done in such a way that they do not degrade performance [28].

2. Resistors

Resistors can be formed from the source–drain diffusions as well as by polysilicon lines. Even though the temperature coefficient, voltage coefficient, and the obtainable ratio accuracy are vastly inferior to MOS capacitors, resistors are sometimes used as precision components. The advantage of resistors is that simpler circuits are sometimes possible, since capacitors cannot supply current and thus cannot act as a voltage source. Also, capacitive dividers require some form of resetting to ensure that parasitic leakage currents do not cause voltage offsets on pure capacitive nodes.

3. MOS Switches

The second principal component in analog MOS circuits is the MOS switch. This device behaves like a resistor in the on state and in the off state like an open circuit. The important parameters of the MOS transistor from the standpoint of an analog circuit component are the on-resistance, off-leakage currents and the parasitic capacitances. Typical NMOS silicon gate technology is capable of producing switch devices with a channel length of 5 μm and, assuming a length/width ratio unity, an on resistance of 5000 Ω (with a gate drive voltage of 5 V with respect to the source). This device would display a leakage current from the source and drain to the substrate of the order of 10^{-14} A at 70°—a value that is negligible in most applications. The parasitic capacitance from source and drain to substrate would be about 0.020 pF each, and the overlap capacitance from drain to gate and source to gate would be about 0.005 pF. The capacitance of the channel when the gate potential is 5 V more positive than the source and drain is approximately 0.008 pF. The values of capacitances used as passive components typically range from 1 to 20 pF. Thus the effect of the various parasitic capacitances on the behavior of the circuit must be carefully considered as well as the flow of the charge stored in the channel of the transistor.

4. Operational Amplifiers

Until very recently, virtually all commercially manufactured operational amplifiers were fabricated utilizing either bipolar technology or a mixture of bipolar and MOS technology. Recently, the trend toward higher levels of integration on MOS–LSI chips has led to the need for all-MOS operational amplifiers to be included on such chips. A considerable amount of recent work has been carried out directed toward the realization of MOS operational amplifiers. Typically achieved levels of performance are a gain of from 60 to 80 dB, common mode and power supply rejection rations of 60 dB, unity-gain bandwidth of 2 MHz, and power dissipation of 1 to 5 mW [29]. These devices typically occupy a die area on the order of 0.2 mm², so that the inclusion of several tens of them on a single chip is feasible.

B. Analog Sampled-Data Filters

The most common signal processing function is frequency selective filtering, which accounts for the variety of methods developed over the years to accomplish this task [28]. The development of low-cost, high-performance, monolithic operational amplifiers led to the utilization of active RC filtering techniques to replace passive RLC filters in the 1960s. These RC active filters have proved difficult to realize in completely monolithic form because stable, precise, RC products cannot be achieved in standard IC processes. As a result, the preferred technological approach to voice-frequency filtering until very recently has been discrete-component or hybrid integrated circuit realizations of various types of active filters. Recent developments in both analog and digital filters using MOS–LSI have made it possible to fully integrate the voice-frequency filtering function, with the usual advantages in cost, size, power, and weight.

While the density advantage of MOS has led to its wide utilization for digital circuits, a second key advantage, which has only recently been recognized, has led to the increasing use of MOS technology to perform analog signal processing. In contrast to bipolar technology, MOS integrated circuits offer the ability to store charge on a node over a period of many milliseconds and to sense the value of the charge continuously and nondestructively. The former results from the high impedance of the MOS transistor in the off state, and the latter results from this essentially infinite input impedance of the MOS transistor. This inherent analog memory capability does not exist in bipolar technology.

The charge storage feature of MOS technology was first utilized in dynamic random access memory design and dynamic logic. In both instances, information storage on small nodal capacitances in the circuit allows the elimination of active or pullup transistors which are otherwise required to preserve the information content of the circuit. This results in greater circuit density. While the information stored on the nodal capacitance is essentially analog in that the node voltage has a continuous range of values, in digital applications the information is usually interpreted as being only one of two logic levels; these binary levels are typically refreshed (restored) every 2 msec.

The charge-storage feature of MOS technology was first used to perform analog signal processing with the advent of the bucket brigade shift register. In this device, a time-varying analog signal is converted to a series of charge packets proportional to the value of the input signal at the sample instants. These charge packets are passed serially through a chain of MOS capacitors, the position of the charge being brought about by the action of appropriately clocked MOS devices. In analog filtering applications of such devices, the charge at each capacitor site is sensed nondestructively by MOS transistors. These sensed voltages can, for example, be added in a weighted summation to produce a single output, providing a realization of a sampled-data transversal filter, as shown in Fig. 13 [30].

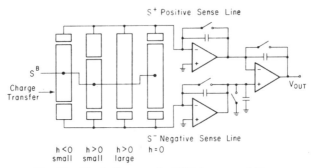

Fig. 13.   Block diagram of a CCD transversal filter.

The charge-coupled device (CCD), which was developed subsequently to the bucket brigade, provides a similar function in analog signal processing applications but with the integration of the storage capacitance with the MOS switch structure and MOS sensing device in such a way that transversal filters and other analog signal processing devices can be fabricated in a smaller silicon area and achieve higher performance. These devices provide an extremely effective technological tool for attacking filtering problems in which transversal filter structures are required, for example, as matched and adaptive filters [31].

However, the utilization of CCD and bucket bridgade filters in large-quantity, low-cost, general filtering applications has been hindered by several factors. First, until very recently, the peripheral circuitry required to convert the input signal into charge packets and to recover a low-impedance voltage signal at the output had to be realized off chip since this circuitry was not thought to be easily realizable on the same chip. Second, the transversal filter itself has certain limitations as a general-purpose analog filtering and signal processing technique. Among these is the fact that since the signal is time sampled at the input to the CCD, a continuous-time antialiasing filter is required to bandlimit the input signal to a frequency below the sampling rate. Since the sampling rate in CCDs must be relatively low (to minimize filter length and conserve silicon area) a relatively complex antialiasing filter is required. Also, since CCD filters are nonrecursive, for transfer functions with a narrow passband and long impulse responses, a very long CCD filter utilizing a large silicon area is required.

In the last several years, the concept of analog charge storage and processing in MOS technology has been applied in new ways and has overcome some of the difficulties inherent in the transversal filter structures. First, in order to obtain highly selective filters with long impulse responses in a small silicon area, one is led to adopt discrete-time, recursive filter structures. A particularly useful class of these filers have proved to be filters that are essentially discrete-time simulations of active filter circuits in the limit of high sampling rates. These filters have come to be called switched-capacitor filters and are distinguished by the fact that they utilize a capacitor and MOS switches to simulate the circuit behavior of a resistor, as shown in Fig. 14 [32].

The operation of this "resistor" is as follows: the switch is initially in the left-hand position so that the capcitor C is charged to the voltage V_1. The switch is then thrown to the right and the capacitor is discharged to the voltage V_2. The amount of charge that flows into (or from) V_2 is thus $Q = C(V_2 - V_1)$. If the switch is thrown back and forth at a clock rate f_c then the average current flow i from V_2 into V_1 will be $C(V_2 - V_1)f_c$. Thus the size of an equivalent resistor that would give the same average current as this circuit is

$$R = 1/Cf_c. \tag{6}$$

If the switched rate is much larger than the signal frequencies of interest, then the time sampling of the signal that occurs in this circuit can be ignored in a first-order analysis and the switched-capacitor can often be considered as a direct replacement for a conventional resistor. If, however, the switch rate and signal frequencies are of the same order, then sampled data techniques are required for analysis and, as for any sampled data system, the input signal should be bandlimited as dictated by the sampling theorem.

The switched-capacitor resistors require very little silicon area to implement large resistance values. In fact, the silicon area decreases as the required value of resistance increases. To implement audio-frequency filters, a resistance on the order of 10 MΩ is needed if a monolithic capacitor of a reasonable size is to be used ($\sim$10 pF). This value of resistance is easily achieved by switching a 1-pF capacitor at a 100-kHz rate, requiring approximately a silicon area of 0.01 mm^2 (to be compared to a total chip area of 10 to 20 mm^2). If a 10-MΩ resistor were implemented by using a polysilicon line or by a diffusion, the area required would be at least 100 times larger.

By using this switched-capacitor resistor in conjunction with other

Fig. 14.   (a) Switched-capacitor equivalent of a resistor; (b) MOS implementation.

capacitors and op amps (operational amplifiers), it is possible to realize many of the circuit configurations used in conventional RC active filters. As in the case of digital filters and CCD filters, the signal in switched-capacitor filters is sampled in time so that antialiasing filtering is required. Fortunately, the penalty in silicon area for a high sample rate is much less in the switched-capacitor filters in comparison to the CCD transversal filters and thus the antialiasing filter can be quite simple.

Viewed from a conventional RC active filter standpoint, these filters have the property, which will be demonstrated later, that all the RC time constants of the filter are determined by the clock frequency that is used to drive the switches and by capacitor ratios. Thus the problem of controlling RC products to a high degree of accuracy, which is characteristic of RC active filter realizations, is reduced for these filters to the problem of maintaining accurate capacitor ratios and supplying an accurate external clock frequency. This property makes the filters much more practical for monolithic realization.

A conventional single-pole RC low-pass filter is shown in Fig. 15a and a switched capacitor implementation of this filter is shown in Fig. 15b. In spite of the simplicity of this filter, it demonstrates some of the most important advantages of the use of switched capacitors. The 3-dB bandwidth of the conventional RC filter is

$$\omega_{3dB}^{RC} = 1/R_1 C_2. \tag{7}$$

The 3-dB width of the switched-capacitor filter can be approximated by substituting the effective resistance of the switched capacitor C_1 into Eq.

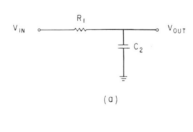

(a)

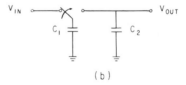

(b)

Fig. 15. (a) Simple RC low-pass filter; (b) switched-capacitor equivalent.

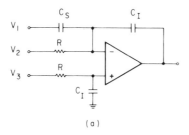

(a)

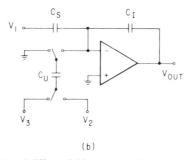

(b)

Fig. 16. (a) Conventional differential integrator; (b) switched-capacitor version.

(6) to obtain

$$\omega_{3dB}^{SC} = f_c(C_1/C_2); \qquad f_c \gg \omega_{3dB}^{SC}. \tag{8}$$

(The constraint that $f_c \gg \omega_{3dB}^{SC}$ is imposed in order that the effects of time sampling and charge sharing can be ignored.) Since the bandwidth of the switched-capacitor filter is proportional to a ratio of capacitor values, it can be accurately defined with a high degree of stability.

From Eq. (8) it can be seen that ω_{3dB}^{SC} is proportional to the clock frequency; thus programmability can be achieved by simply varying the clock rate (a property that is common to all discrete-time filters).

The problem of implementing active filters in MOS technology is thus reduced to the question of what kind of active filter should be used. The filter configuration chosen is the analog computer simulation of the equations that describe a passive, doubly terminated, RLC ladder [33,34]. These filters are called "leap frog" or "active ladder" filters in the active filter circuit literature and are closely related to wave digital filters in the digital signal processing literature.

The basic building block of these filters is a differential integrator and summer, shown in Fig. 16a, that is implemented in the conventional way with resistors and capacitors. The equation that describes this circuit is

$$V_{out}(\omega) = -\frac{C_s}{C_I} V_1 + \frac{1/RC_I}{j\omega} (V_3 - V_2). \tag{9}$$

Straightforward substitution of the resistors by the circuit in Fig. 14 would yield a switched-capacitor circuit that is sensitive to parasitic capacitances, so an alternate method of realizing the resistors which also makes it possible to invert the sign of the signal is used and is shown in Fig. 16b [35]. The circuit in this figure is described by the z-transform equation

$$V_{out}(z) = -\frac{C_S}{C_I} V_1 + \frac{C_U/C_I}{1 - z^{-1}} (z^{-1}V_3 - V_2). \tag{10}$$

Appropriate interconnection of these sampled-data integrators following an RC active filter prototype yields the parasitic free, fifth-order, switched-capacitor ladder shown in Fig. 17. A microphotograph of a ladder filter is shown in Fig. 18, in which the ratioed capacitors, which are formed by connecting unit capacitor squares, are clearly visible.

C. Digital-to-Analog Converters

The two basic techniques that have emerged as the most suitable for precision D/A converters, which are compatible with standard MOS technology, are voltage division using a resistor string and charge redistribution using MOS capacitors [36].

1. Resistive Voltage Dividers

The most straightforward form of D/A conversion (but not the most accurate) is to divide a reference voltage into the required number of quantized voltage levels by using a string of 2^m series resistors for an m-bit

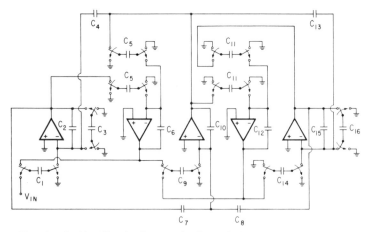

Fig. 17. Ladder filter implemented using switched-capacitor integrators.

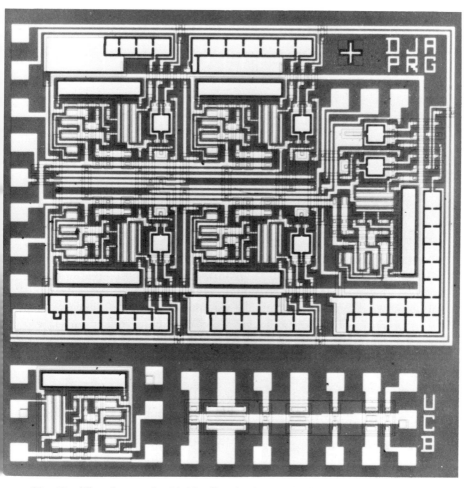

Fig. 18. Microphotograph of ladder filter implemented using switched-capacitor techniques.

converter, as shown in Fig. 19 [37]. A 1 of 2^m decoding tree, which is shown in the figure, can be used to tap into the desired point on the resistor string. This tree is particularly compact in MOS technology because of the availability of the MOS switch, which can be merged with the resistor string if the resistor string is fabricated using the source–drain diffusion. The area for an 8-bit converter is quite small, and 10- and 12-bit converters have been fabricated with some slight modifications of this basic technique.

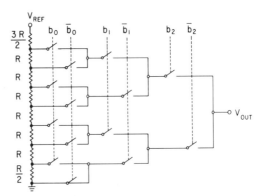

Fig. 19. Resistor string digital-to-analog converter.

2. Capacitive Charge Redistribution

Voltage division can also be obtained with the use of capacitors as the passive elements, as in Fig. 20 [38]. The operation of this converter begins by a resetting of the top plate to ground. Following this the desired combination of switches on the bottom plate are thrown to the reference voltage V_{ref}. A charge redistribution occurs which results in voltage on the top plate of

$$V_{out} = V_{ref} \sum_i b_i 2^{-i}, \qquad (11)$$

where b_i is one bit of the digital word being converted. Multiplying D/A converters have been fabricated using this technique. These circuits have proved to be very efficient for performing multiplications in hybrid analog/digital signal processing systems [39].

D. Analog-to-Digital Converters

A variety of methods has been developed to perform A/D conversions, which trade off accuracy, speed, and circuit area. The basic technique of all A/D converters is that a precise voltage is generated by a D/A converter, which is then compared to samples of the incoming analog signal. Since the incoming signal must be sampled in time, it has to be band-limited before conversion. The three basic types of converters are serial, successive approximation, and flash (or parallel) [36].

1. Serial Converters

This type of converter is for signals with low bandwidths (<100 Hz) and high accuracy (>12 bits). An example of a serial converter is shown in

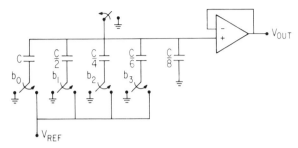

Fig. 20. Circuit of a capacitive redistribution digital-to-analog converter.

Fig. 21. Switch 1 connects the integrator within the dashed lines to either the magnitude of the sampled input signal $|V_{in}|$ or to a negative reference voltage $-V_{ref}$. The conversion begins by resetting the op amp integrator with switch S_2 and integrating until the comparator is at the threshold of switching V_{th}. The integration is performed using the same technique as the switched-capacitor filters in which the capacitor C_0 is switched between the op amp and ground. The integrator is then switched another fixed number of clock cycles N_{ref}. The voltage V_{Int} at the output of the integrator is then

$$V_{int} = (C_0/C_I)N_{ref}V_{in} + V_{th}.$$ (12)

The next step is to integrate the negative reference voltage until the integrator output is again at the comparator threshold. The number of clock cycles (i.e., times C_0 is switched), which is measured by the counter N_{out}, is given by

$$N_{out} = (V_{irt} - V_{th})/(C_0/C_I).$$ (13)

Combining Eqs. (12) and (13) and assuming that the reference count

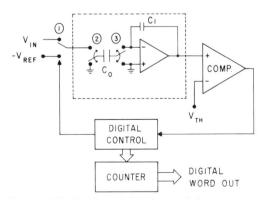

Fig. 21. Circuit of a serial analog-to-digital converter.

$N_{ref} = 2^m$, where m is the number of bits desired, N_{out} can be written as

$$N_{out} = 2^m V_{in}/V_{ref}, \qquad (14)$$

which is the desired output digital word. Since N_{out} is not a function of the comparator threshold, the absolute value of any of the capacitors, or the clock rate, extremely high accuracies are potentially possible. The limitations on accuracy are charge injection by the switches and noise level of the comparator. Since it takes 2^m clock cycles for an m-bit conversion, this method is inherently slow, but the small amount of analog circuitry should make serial approaches more useful as the speed of the technology is increased.

2. Successive-Approximation Converters

The use of successive approximation is the most important conversion technique for signal processing, since it can convert signals with high accuracy at bandwidths up to the present limits of MOS digital processors. In Fig. 22 is shown a block diagram of a successive-approximation A/D converter. The D/A converter in this figure can be implemented in a variety of ways (i.e., with resistors or capacitors), as long as a discrete voltage level can be generated in a few clock cycles. In the serial converter, since each voltage is generated in sequence, the tests (comparisons of the input signal with the discrete voltage levels) must also be sequentially performed. In a successive-approximation converter a carefully selected sequence of tests is performed, which allows rapid determination of the proper digital output word. The successive-approximation sequence requires m clock cycles for an m-bit conversion. The disadvantage is that the resolution of this type of converter is determined by the absolute accuracy of the D/A converter.

The first test performed (after the sign of the signal is found) determines if the magnitude of the input signal lies in the upper or lower half of the allowed input range, i.e., is between 0 and $V_{ref}/2$ or between $V_{ref}/2$ and V_{ref}. This yields the most significant bit of the digital output word. The speed advantage is obtained, because if the voltage is found to be in the lower

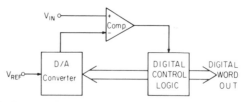

Fig. 22. Block diagram of a successive approximation analog-to-digital converter.

half, none of the voltages in the upper half need be tested (and vice versa). Next, a test is performed to see if the input is in the upper or lower half of the remaining allowed range to determine the next most significant bit. Again, the result of this test allows the elimination of the testing of half of the remaining voltage levels. This sequence of tests continues until the least significant bit is found.

The bandwidth of signals that can be processed by successive-approximation converters ranges up to several megahertz and accuracies as high as 12 bits have been obtained [40,41]. There are presently a number of efforts that attempt to obtain monolithic converters in the range of 14 to 16-bit accuracies for use in consumer digital audio applications.

3. Flash Converters

Through the use of parallelism a significant increase in bandwidth can be obtained at the expense of increased silicon area. The flash (or parallel) conversion approach is shown in Fig. 23. In this technique there are m comparators for an m-bit conversion. The resistor string can be formed by tapping a polysilicon line or a diffusion in the silicon. The conversion time for this approach is just the time it takes for a single converter to make a comparison and thus the data rate can be extremely fast. However, since the amount of circuitry is increasing exponentially with number of bits, the maximum accuracy at present device densities has been limited to about 6 bits (64 comparators) at 20 MHz [42]. This type of converter should be able to easily exploit the increased capabilities of VLSI technology and thus will probably be of increased importance as the technology is scaled.

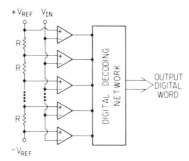

Fig. 23. Block diagram of a flash (parallel) A/D converter.

E. Analog Circuit Scaling

The analog circuits described in the last few sections have proved to be very useful in today's technology. However, in order to assess their future importance, the effect of scaling on their performance must be investigated. The patterning process used for the passive elements is the same process that defines the features of the digital logic gates. As the device dimensions are scaled, this patterning process must be improved. It will, therefore, be possible to obtain improved ratio accuracies of passive components (MOS capacitors and resistors) if their size is not scaled down or the same accuracy will be achievable in less area. Analog amplifiers will also be able to take advantage of the speed improvement obtained by the decreased capacitances of a scaled technology.

The major problem with scaling the analog circuits is the effect on the dynamic range, which is the ratio of the maximum signal that can be processed to the noise level. Since the maximum signal level decreases linearly with the supply voltage, if equivalent reductions are not obtained in the noise level when the supply is decreased, then the dynamic range will be reduced. Fortunately, at this time there is sufficient margin that the supply can be substantially decreased over present-day voltages (10 V) and still retain adequate dynamic range for many applications.

To demonstrate this point, it is first necessary to enumerate the fundamental noise sources and determine how they scale. There are two sources of the random noise in an MOS circuit: $1/f$, or surface state noise, and thermal noise. The surface state noise has a power spectral density that increases as the frequency is reduced and is thus more important at low frequencies (<10 kHz). However, with some increase in circuitry, it is possible to essentially eliminate this noise source by signal-chopping techniques such as chopper stabilization and correlated double sampling, which significantly reduce the noise at frequencies below the chopping rate [43,44]. This is fortunate since the $1/f$ noise component increases as the gate capacitance is decreased.

The contribution of the thermal noise to the total circuit noise of some circuits (e.g., the switched-capacitor filters) increases as the technology is scaled, since the inherent bandlimiting by the circuit capacitances is reduced, if the standard scaling rules optimized for digital circuits are slavishly followed. This occurs in spite of the fact that the thermal noise itself (which is proportional to the square root of the channel resistance) is actually decreasing as $K^{1/2}$, where K is the scaling parameter (see Section III). However, if the capacitor areas are not scaled, the value of the capacitors will increase as the scaling factor, because the insulator thickness is being reduced, and the total thermal noise contribution will actually decrease as $K^{1/2}$. This noise decrease is, of course, at the expense of reduced circuit density, so that a trade-off between silicon area and dynamic range will be required for a given application. An estimate of the reduction of the dynamic range can be made by assuming that the maximum signal decreases with the the supply voltage which scales down as K and yields a decrease in dynamic range by the square root of the scaling factor $K^{1/2}$. If it is assumed that between now and 1990 the scaling will be by a factor of 5 (i.e., supply voltages that are now 10 V for analog circuits will be reduced to 2 V), the reduction in dynamic range will be the very moderate level of 7 dB. At this time complex analog filtering circuits (~ 20 poles) are being fabricated with dynamic ranges on the order of 97 dB [45]. A reduction in dynamic range to 90 dB remains more than adequate for most signal processing applications.

VI. CONCLUSIONS

By using the fact that VLSI is just the continuing development of MOS technology, it is possible to extrapolate the trends of the past 10 yr to obtain predictions about the capability of MOS technology in the next 10 yr. Some of the implications of these predictions were presented. It was shown that even at the limits of scaling the processing bandwidth of MOS technology will only have improved to the point where it is equivalent to today's advanced bipolar technologies. Because of the increased capital costs of implementing VLSI technology, the access to the technology will become increasingly limited. The result of this trend is that the development of VLSI signal processing systems will be by large vertically integrated system/IC manufacturers. Because the VLSI signal processing chips will be able to implement significant system functions, it will be necessary for the system designer to be involved in the chip design. This will be made possible by standardized design methods such as gate arrays, standard cells, and PLAs, which will allow system designers to design the VLSI chips themselves. A variety of analog design techniques have been developed that are compatible with MOS technology and it is expected that these techniques will continue to be important as the technology is scaled. In fact, if progress is not made in the analog interface circuits that is equivalent to the improvements in the digital processors, then the ability for VLSI to make an impact will be greatly diminished. In spite of the changes that VLSI will require, the high level of system integration that it will make possible will result in widespread use of sophisticated signal processing algorithms because of the reductions in system cost, size, and power.

REFERENCES

1. S. Waser, Survey of VLSI for digital signal processing, *Proc. ICASSP, Denver Colorado* pp. 376–379 (1980).
2. J. F. Bucy, Texas Instruments' First Quarter and Stockholders' Meeting Report, p. 12 (1980).
3. D. A. Hodges, P. R. Gray, and R. W. Brodersen, Potential of MOS technologies for analog integrated circuits, *IEEE J. Solid State Circuits* SC-13, 285–294 (1978).
4. M. Townsend *et al.*, An NMOS microprocessor for analog signal processing, *IEEE J. Solid State Circuits* SC-15, 33–38 (1980).
5. R. W. Blasco VMOS chip joins microprocessor to handle signals in real time, *Electronics* 52, No. 18 (1979).
6. J. R. Boddie *et al.*, A digital signal processor for telecommunications applications, *Int. Solid State Circuits Conf. Tech. Digest, San Francisco, California* pp. 44–45 (1980).
7. Y. Kawakami *et al.*, A single chip signal processor for voiceband applications, *Int. Solid State Circuits Conf. Tech. Digest, San Francisco, California* pp. 40–41 (1980).
8. R. Wiggens and L. Brantingham, Three-chip system synthesizes human speech, *Electronics* 51, 109–116 (1978).
9. Y. Chen and D. L. Duttweiler, A 35,000 transistor VLSI echo canceller, *Electronics* 51, 42–43 (1978).
10. F. I. Itakura and S. Saito, Analysis–synthesis telephony based on the maximum likelihood method, *Proc. Int. Congr. Acoust., 6th Tokyo*, pp. 17–20 (1968).
11. I. A. Young, D. B. Hildebrand, and C. B. Johnson, A low power NMOS transmit/receive IC filter for PCM telephony, *Int. Solid State Circuits Conf. Tech. Digest, San Francisco, California* pp. 184–185 (1980).
12. B. J. White, G. M. Jacobs, and G. F. Landsburg, A monolithic dual tone multifrequency receiver, *IEEE J. Solid-State Circuits* SC-14, 991–997 (1979).
13. G. Moore, VLSI: Some fundamental challenges, *IEEE Spectrum*, 30–37 (1979).
14. R. P. Capece, Special report: New LSI processes, *Electronics* 52, 109–115 (1979).
15. H. N. Yu *et al.*, 1 micron MOSFET VLSI technology: Part 1, *IEEE Trans Electron Devices* ED-26, 318–324 (1979).
16. R. M. Jecmen *et al.*, HMOS II static RAMs, *Electronics* 52, 124–128 (1979).
17. L. M. Dang, A simple current model for short channel IGFET's, *IEEE Trans. Electron Devices* ED-26, 436–445 (1979).
18. P. A. Hart, T. V. Hof, and F. M. Klassen, Device down scaling and expected circuit performance, *IEEE Trans. Electron Devices* ED-26, 421–429 (1979).
19. M. T. Elliot *et al.*, Size effects in e-beam fabricated MOS devices, *IEEE Trans. Electron Devices* ED-26, 469–475 (1979).
20. T. Yamaguchi, M. L. Lust, S. Ragsdale, and S. Sato, A new submicron channel/high speed MOS–LSI technology, *IEEE Trans. Electron Devices* ED-26, 611–622 (1979).
21. B. Hoeneisen and C. A. Mead, Fundamental limitations in microelectronics—I. MOS technology, *Solid-State Technol.* 15, 819–829 (1972).
22. R. W. Keyes, The evolution of Digital Electronics towards VLSI, *IEEE Trans. Electron Devices* ED-26, 271–279 (1979).
23. B. LeBoss, Speech I/O is making itself heard, *Electronics* 53, 95–105 (1980).
24. A. Weinberger, High speed PLA adders, *IBM J. Res. Dev.* 23, No. 2, 163–178 (1979).
25. K. Hwang, "Computer Arithmetic." Wiley, New York, 1979.
26. N. Bernstein, What to look for in input/output boards, *Electronics* 51, 113–119 (1978).
27. J. L. McCreary and P. R. Gray, All-MOS charge redistribution analog–digital conversion techniques—Part 1, *IEEE J. Solid-State Circuits* SC-10, 371–379 (1975).

28. R. W. Brodersen, P. R. Gray, and D. A. Hodges, MOS switched-capacitor ladder filters, *Proc. IEEE* **67**, 61–75 (1979).

29. D. Senderowicz, D. A. Hodges, and P. R. Gray, A high performance NMOS operational amplifier, *IEEE J. Solid-State Circuits* **SC-13**, 760–766 (1978).

30. D. D. Buss, D. R. Collins, W. H. Bailey, and C. R. Reeves, Transversal filtering using charged coupled devices, *IEEE J. Solid-State Circuits* **SC-8**, 138–146 (1973).

31. C. R. Hewes, R. W. Brodersen, and D. D. Buss, *Proc. IEEE* **67**, 1403–1415 (1979).

32. D. L. Fried, Analog sampled-data filters, *IEEE J. Solid-State Circuits* **SC-7**, 302–304 (1972).

33. B. J. Hosticka, R. W. Brodersen, and P. R. Gray, MOS sampled data recursive filters using switched-capacitor integrators, *IEEE J. Solid State Circuits.* **SC-12**, 600–608 (1977).

34. J. T. Caves *et al.*, Sampled analog filtering using switched-capacitors as resistor equivalents, *IEEE J. Solid-State Circuits* **SC-12**, 592–599 (1977).

35. G. M. Jacobs, D. J. Allstot, R. W. Brodersen, and P. R. Gray, Design techniques for switched-capacitor ladder filters, *IEEE Trans. Circuits Syst.* **CAS-25**, 1014–1021 (1978).

36. P. R. Gray and D. A. Hodges, All-MOS analog–digital converter techniques, *IEEE Trans. Circuits Syst.* **CAS-25**, 482–489 (1978).

37. A. R. Hamade and E. Campbell, A single-chip 8-bit A/D converter, ISSCC Digest Tech. Papers, pp. 154–155 (1976).

38. J. F. Albarron, A charge-transfer multiplying D to A converter, *IEEE J. Solid-State Circuits* **SC-11**, 772–779 (1976).

39. R. D. Fellman, P. J. Hurst, and R. W. Brodersen, MOS analog/digital techniques for linear prediction, *Proc. Int. Conf. Communications, Seattle,* pp. 100–105 (1980).

40. B. Fotouhi and D. A. Hodges, High resolution A/D conversion in MOS–LSI, *IEEE J. Solid-State Circuits* **SC-14**, 920–925 (1979).

41. J. J. Connolly, T. O. Redfern, S. W. Chin, and T. M. Frederiksen, A monolithic 12-bit + sign successive proximation A/D converter, *Int. Solid-State Circuits Conf. Digest Tech. Papers,* pp. 12–13 (1980).

42. A. G. F. Dingwall, Monolithic expandable 6 bit 20 MHz CMOS/SOS A/D converter, *IEEE J. Solid-State Circuits* **SC-14**, 926–937 (1979).

43. R. Poujois, J. M. Ittel, and J. Borel, A low drift MOSFET operational amplifier ARZ, *Eur. Solid-State Circuits Conf. Digest Tech. Papers* pp. 14–15 (1976).

44. K. C. Hsieh and P. R. Gray, A low-noise chopper-stabilized differential switched-capacitor filtering technique, *Int. Solid-State Circuits Conf. Digest Tech. Papers,* pp. 128–129 (1981).

45. W. C. Black, D. J. Allstot, S. H. Patel, and J. B. Wieser, CMOS PCM channel filter, *Int. Solid-State Circuits Conf. Digest Tech. Papers* pp. 84–85 (1980).

Chapter **8**

The Impact of VLSI on Military Systems: A Supplier's Viewpoint

DENNIS R. BEST

Microelectronic Circuit Development Laboratory
Equipment Group, Texas Instruments, Incorporated
Dallas, Texas

I. INTRODUCTION—STANDING AT THE THRESHOLD

A. Perspective

As we stand on the threshold of VLSI, it would appear that the exploitation of this technology for national defense systems would offer limitless

229

capabilities restricted only by the prognosticator's imagination. However, this view is obscured by the realization that today's LSI technology has not achieved, on military electronics equipment, the impact that it has had in the commercial sector.

Therefore, to make realistic projections of the impact of VLSI on the military, we must examine this technology in light of the total military electronics environment. This view must include the events and structure of electronics and the military marketplace in the past and consideration of the technical, economic, and political pressures that are creating the environment of the future.

We must also examine this nation's strategic and tactical battlefield and consider the potential that VLSI may have on our nation's defense objectives and systems and project this potential into the trends in the electronics content of the Department of Defense (DoD) procurement and research and development (R&D) budgets and activities, including the integrated circuit content.

It will also be useful to examine the DoD's thrust to overcome the dilemma of inserting the enhanced performance inherent in VLSI into high-impact military equipment while achieving significant life cycle cost reductions: the six-year, $200 million Very High Speed Integrated Circuits (VHSIC) program. We can then conclude by projecting the technologies, components, and system environments of the VLSI era.

Throughout, we shall attempt to point out our view of the ultimate payoff of VLSI for the military—affordable, effective "force multiplication" for our national defense.

B. DoD and Electronics—The Past

Military systems requirements for high-performance digital technology, with the stringent size, power, weight, temperature, and reliability constraints associated with such systems, provided the impetus that led to the invention of the digital integrated circuit and then the funding, through military R&D programs, that created the digital integrated circuit market as we, in both the commercial and military electronics sectors, know it today. In the late 1950s, during the age of the transistor, perceptive military laboratories were funding the emerging semiconductor companies' independent research and development and manufacturing methods programs that eventually led to the successful reduction to practice of the integrated circuit. Then, in the early 1960s, military system program office's acceptance of the integrated circuit endorsed the viability of the integrated circuit concepts for military equipment.

In the late 1950s, digital technology, implemented with discrete transistor and passive components, was a dominant factor in the development and production of large military systems. The military laboratories' research and development funding was concentrating on the packaging of discrete logic to fit the needs of the newer, more complex systems. The Army Signal Corps had a $25 million program for developing microminiature, common stackable components—the Micro-Module program. This contract, primarily through RCA, was to create a family of logic components conforming to a defined physical standard for use throughout the military. The Navy did not have a program, but the Navy laboratories were favoring thin-film or hybrid packaging techniques for their logic requirements. The Air Force had a $2 million/yr research and development effort, primarily with Westinghouse, aimed at a concept called Molecular Electronics. This approach called for a departure from conventional electronic circuits in order to develop new structures to perform the desired functions in new ways. For example, a quartz crystal would be preferable to inductors and capacitors, without a part-for-part equivalence of the past electronic methods. This approach was, however, considered too impractical for use by most service laboratories.

Shortly after the public disclosure of Jack Kilby's "solid circuit" (integrated circuit) concept at the March 1959 IRE show, Texas Instruments began to promote the concept within the military. Within the Air Force a small group, led by R. D. Alberts of the Wright Air Development Center, under the loosely interpreted umbrella of the molecular electronics program, provided a series of research and development contracts to broaden the integrated circuit concept. The laboratory also dedicated manufacturing methods funding to support the installation of the first integrated circuit manufacturing line. These manufacturing methods programs included demonstration vehicles that would clearly show the advantage of integrated circuits for the military.

By the end of 1961, Texas Instruments had delivered the "molecular electronics computer" to the Air Force and had announced the Series 51 digital logic family. During the same year, Fairchild had introduced their micrologic family. However, digital IC technology was still a novelty. The molecular computer was basically a 10-bit, 200-kHz clock calculator with keyboard and display, but it did have a very aggressive packaging technique, as shown in Fig. 1. The computer contained 587 circuits packaged in stacked IC modules with the most complex integrated circuit being a single flip-flop. The electronics dissipated 16 W of power and was packaged in a 10-oz, 6.3-in^3 container.

In 1962 two events took place that were to move the digital integrated circuit from novelty to practical utilization. Autonetics awarded Texas In-

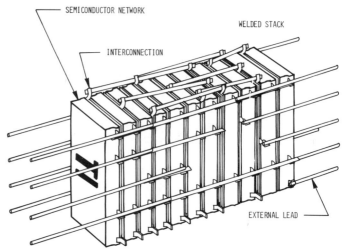

Fig. 1. Molecular electronics computer IC module packaging.

struments a large contract for the design and development of twenty-two special integrated circuits for the Minuteman II missile, and Fairchild received a contract from NASA to develop the Apollo computer. These two programs endorsed the validity of the use of integrated circuits in the military environment. In particular, since the programs required the extreme of high reliability, the door was now opened for the use of ICs by all military programs.

These events, in effect, provided the trigger for the explosive growth of the military market for digital integrated circuits. In 1964 the Series 54 Transistor–Transistor Logic (TTL) logic was announced. This family, although extended and much improved over the years, primarily in speed–power performance, is still the most widely used military technology of today. This market growth is illustrated in Fig. 2. Until the later 1960s, the IC market was as much as 98% military—the very large commercial systems houses, such as IBM and Bell, were still dedicated to hybrid circuits based on the premise that you could indeed make "better" transistors and passive components discretely at a lower cost. The first integrated circuits were priced at $100 each in small volume, and the combining of resistors and capacitors in one monolithic circuit with a transistor certainly degraded the ultimate performance of that transistor.

However, in the late 1960s, certain smaller companies, such as Digital Equipment Corporation (DEC), were quick to recognize the value of single-packaged logic "functions." Very rapidly, DEC integrated ICs into their "logic module" business, and, by 1970, they, along with others, had introduced the first of the minicomputers—a new electronics business that was the first of many to create an IC market force counter to the purely military requirements.

By the early 1970s, the major computer and communication system houses had boarded the integrated circuit bandwagon and developments for either the commercial or military market sectors benefited both sectors. The dominant technology remained TTL, which inherently met the unique temperature, radiation, and reliability requirements of the military. Although the Emitter Coupled Logic (ECL) integrated circuit generated considerable interest in this time period, its higher power and thus more costly packaging requirements in the military system platforms constrained its applicability and thus its total market and development emphasis.

However, the growth of the commercial IC market, while offering the military a wider range of semiconductor products at ever lower cost, was slowly eroding the military's dominant role in the direction of IC developments. Then the event occurred that ended the peaceful coexistence of the military and commercial markets—the invention of the single-transistor dynamic metal–oxide–semiconductor random access memory (MOS RAM) cell.

C. DoD and Electronics—The Present

The single-cell dynamic random access memory (DRAM) permitted the production of the first truly cost-effective solid-state memory device that had sufficient density to allow combining address decoding logic on the same chip with memory cells. Progress had been made in solid-state memory prior to the DRAM, as indicated in Fig. 3, but this step function improvement created a major redirection of the semiconductor market. In

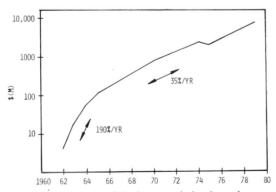

Fig. 2. Growth of the integrated circuit market.

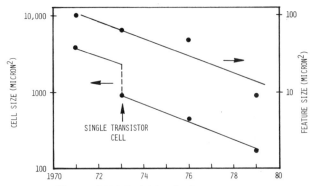

Fig. 3.   MOS RAM density improvements.

fact, totally new markets were to be created, primarily in the consumer arena, and many new semiconductor companies were formed to take advantage of this new, low-cost technology serving new markets that were free of the stringent military procurement, environment, and quality requirements. Even the major computer markets shifted emphasis—the freedom from the constraining size and cost of core memories allowed a more liberal use of memory in computer and logic structures, resulting in radical innovations in the main-frame and minicomputer markets. Today, memory accounts for almost one-half of the total sales of LSI devices and is expected to gain on that percentage in the future.

By 1975 the attention to MOS technology had extended from memory to logic and the first one-chip calculator had been introduced. One device, selling for around $2, had replaced the 89 chips with a total value of approximately $170 required to perform the calculator function in 1967.

The higher return on investment for semiconductor companies in the commercial arena redirected their developments toward the computer, industrial, and consumer market segments. As the high volumes associated with a single product brought IC prices down the traditional 70% semiconductor learning curve, the commercial markets expanded exponentially. Although the military benefited indirectly from the increased electronics market, a structural change was occurring in the military semiconductor market structure.

As commodity suppliers became increasingly disinterested in the military market, with the military segment going from 70% of the total semiconductor market in 1965 to the 7–10% estimate for 1979, their R&D dollars went into commercial development and they withdrew from heavy participation in the government-funded programs initiated through the DoD laboratories. Their alliances with the laboratories were replaced by the traditional military *systems* contractors who, frustrated in their own attempts to obtain uniquely military ICs from commodity semiconductor suppliers, internally developed an IC R&D laboratory and/or limited-production facilities. These contractors, with a vested interest in military equipment, were more willing to work on the uniquely military performance and environment problems. Where the commodity suppliers, even those with a considerable share of the military market, were unwilling to invest in the technologies, the special processing and testing procedures, or the low-volume design and manufacturing runs necessary to support the military, the systems houses attempted to fill the gap between need and availability.

However, the vertically integrated systems companies still have not solved the puzzle of the use of LSI in military equipment. An excellent example is illustrated by the history of the development of the complementary metal–oxide–semiconductor/silicon-on-sapphire (CMOS/SOS). In theory and in the R&D laboratories, this technology appeared to offer the perfect military LSI solution. In fact, approximately \$20 million/yr of government R&D funding has been dedicated to making this technology a reality in military equipment. However, it has yet to make inroads into any significant piece of military hardware or to reach a performance in production that is significantly better than the commercial mainstream technology devices.

By 1980 the military systems houses, although spending much DoD contract money and their own independent research and development money on alternate IC technology and circuits, were still returning to the commodity semiconductor suppliers for ICs to use in production equipment. Bipolar, for example, TTL, ECL, and, more recently, I^2L (integrated injection logic), remains the dominant military technology for logic and processors, although the N-channel metal–oxide–semiconductor device (NMOS) has reached 60% of the total market for memory devices. Many times the devices are nearing obsolescence when they start their life in the final product. Military systems would often go from concept to operation in five years in the 1950s but take from seven to fifteen years in 1980, whereas in the commercial sector complex components for which no current markets exist often create their own markets and enter volume production of end-user equipment in as little as two years.

If technology has simplified the problems of system design through integration, why are military equipment acquisition cycles lengthening? And, in spite of the high degree of attention to the use of reliable components in military systems, why do the radar and communication equipment used by commercial airlines achieve an 800-hr mean time between failures (MTBF) when the military versions, often 10 times or more as expensive, average only 5–10 hr between breakdowns? Why has the rush to vertical integration by systems suppliers not been able to successfully adapt the functional performance and reliability of LSI into the military segment?

The reasons for the disappointingly slow exploitation of LSI into military equipment are many and varied. The military is a low-volume market for any single device type, while the merchant semiconductor suppliers have concentrated on high-volume devices. The primary military requirement is for high-performance (high throughput at acceptable power dissipation) IC memory and logic functions for unique signal and data processing systems having no commercial equivalent, while the commercial emphasis is on reduced costs through increased functional density, i.e., cost-effectiveness not performance. The military requires the full $-55-125°C$ temperature range, a high component reliability level verified by extensive test procedures, and some degree of radiation hardness, while the commercial market is satisfied with the $0-70°C$ temperature range, the reliability improvements attainable with reduced part counts, and little attention being given to radiation tolerance.

Because the merchant suppliers have not met the military requirements, the military systems suppliers have resorted to dedicated custom IC designs, often with custom IC technologies not aligned to mainstream semiconductor industry capabilities, or the use of commercial devices screened for military operating environments. This approach has been partially responsible for the even higher component and system development and acquisition costs and increased development cycles, as well as increases in the logistics and maintenance, or cost of ownership, of military systems. The long payback period after the investment in technology and equipment to supply military ICs and the higher processing, testing, manufacturing, and documenting costs of the market serve to compound the problems.

For VLSI to have a major impact on military equipment capabilities, we must apply the formula that has lead to the spectacular impact of LSI in the commercial sector as well as recognize the potential for innovation available with VLSI that is neither practical nor apparent with LSI.

D. DoD and Electronics—The Future

The VLSI era, which is better defined as the micron and submicron domain, will have the inherent benefit (gate speed and gate density) of allowing the military supplier to overcome the problems in the utilization of LSI if we successfully manage the application of this technology. The key is to use the silicon area of integrated circuits to reduce the *apparent* complexity of military systems. Military systems are, in fact, becoming more complex but we must learn to manage this complexity and thus control the life cycle costs of military equipment.

The primary utilization of silicon area is to increase the functional density of integrated circuit devices as we go from today's standard 5-μm semiconductor processing to the 1-μm era. However, the models applied to extrapolate the expected speed–power performance and density at 1μm are different for different technologies and circuit structure (logic, ROM, and RAM). The modeling and experimental work being done in semiconductor laboratories today indicates that we can achieve a 3–10 × improvement in logic speed while limiting the maximum power dissipated in one IC to less than 1 W. Logic density will increase from the 5000–7000 gates of today to the 15,000 to 25,000 gates/chip of tomorrow, primarily constrained by power dissipation. One to 2 nsec in-place logic delays will allow the construction of microcomputers executing at 20 to 30-MHz clock rates with 15,000 to 20,000 logic gates and 100,000 ROM bits/chip.

Some of this increased functional density will be used to attack a problem critical to IC manufacturers, systems manufacturers, and end users of electronic equipment: the testability problem. More attention will be given to design-for-test considerations at the system and device level. On-chip logic and memory will be devoted to providing built-in test aids and a limited form of self-diagnosis and perhaps even a limited self-healing capability. Examples of self-healing include the use of error correcting codes (ECC) for memories or perhaps programmable logic arrays (PLAs) for mapping out defective bits in a large memory with redundant cells. However, it should be noted that basic IC reliability will still be a function of the quality of the materials and processes used in the manufacture of ICs, but system reliability, which is a function of the reliability of the ICs, the number of ICs, and the number of interconnections, should see much improvement for a given level of complexity.

Another way of controlling life cycle costs is to use some of the functional density of VLSI to reduce the design costs of ICs or systems. An example of this is the higher-level language structures being built into today's computers to make them easier to understand and program. However, we do not believe that this will be the case for most military equipment designs. End equipment for the military falls into two basic categories: (1) small volumes of very high performance complex systems, such as airborne radars and intelligence centers, or (2) large volumes of low-cost, expendable weapons and man-portable communication equipment. For this equipment, minimizing production costs or maximizing performance will weigh more heavily than design cost.

There are two basic techniques being promoted by industry to ensure rapid integration of VLSI into military equipment. One is the "silicon foundry'' approach. This approach implies the development of design automation techniques that would allow the rapid, low-cost design and manufacture of custom VLSICs. The system developer would have responsibility for generating the data base defining the IC structure and tests, and the semiconductor manufacturer would have the automated manufacturing equipment to quickly produce and deliver the IC. This approach is questionable since historically, integrated circuit complexity has increased at a rate of 1.5 to $2\times$ the rate of design cost reduction per gate attained by design automation techniques. Also, this will not reduce the proliferation of custom ICs in the DoD's logistic inventory. In addition, the same investment in IC manufacturing equipment can provide $10\times$ the output of high-volume, single-design ICs compared to a large number of low-volume IC part types.

The second technique is for the semiconductor manufacturers to provide programmable system components (PSCs) and the data sheets and support software necessary to program the devices to the end-use application. This allows the definition of circuits that are broadly applicable and cost-effective in a broad range of military applications, just as the microprocessor has done in the commercial market. This approach appears to be the most rational since it retains the production volumes traditionally required to achieve cost and reliability goals in IC manufacturing but allows for customization of the circuit to the application via ROM, RAM or logic array programming. It is also relevant to note that the relatively new use of microprocessor support software development tools by system developers has resulted in a 90% learning curve reduction in the cost to program the devices for a given application. It is expected that the ultimate market structure of the VLSI era will have elements of both approaches—programmable system components for the majority of system applications plus a rapid turnaround logic array capability where needed to solve unique systems problems or where the volume of systems allows a more custom approach.

In order to achieve truly step function improvements in military equipment capability, we must do more than just take advantage of the ''shrink'' or further integration of existing system concepts by the use of VLSI. We must promote innovation in the development, procurement, and production of the new technological products. This added technological capability will allow the designer new latitudes of freedom to generate new system, processor, and device architectures. We must be careful not to inhibit the new approaches by the pressures imposed by the problems with utilization of today's technology. Rapidly escalating design and maintenance costs are pushing the military more into the standardization world in both hardware and software structures, but standardization can only occur after the architecture, I/O structure, or process technology matures. This standardization can constrain us from system improvements through technological insertion beyond that available with technology shrinks.

It is already apparent that new alliances will form among the military, the systems companies, and the semiconductor suppliers. We can no longer segment the system and application from the component. The commercial microcomputer experience has shown that the definition of the product and its application support tools are more costly but more important to its ultimate effectiveness than the design of that component, yet the production expertise of the semiconductor manufacturers is key to obtaining the reliability and performance achievable within a given technology.

The impact of the successful utilization of VLSI for national defense requirements can be truly spectacular. Consider the rather simple example of battery costs for a man-portable communications module, for which the government currently spends \$140/W annually. If the application of VLSI can reduce the power consumption of the radio from 10 to 1 W and if we assume a total deployment of 50,000 units, then a \$63 million annual savings would result. On the other hand, the system size reduction available through the use of VLSI could potentially be used to reduce the diameter of defense suppression missiles from 12 to 16 in. to the desirable 4-in. diameter, which, along with the resulting weight reduction, could greatly extend the range of the missiles as well as allow the weapon platform to carry more of the missiles. These savings can be achieved by simply applying higher levels of integration to current weapons concepts.

The enhanced size, power, and performance of VLSI systems may offer the signal processing technology required to take advantage of the total information bandwidth already available from current radar, infrared, and passive radio-frequency sensors and communications transmitter–receivers. This may allow the implementation of high survivability concepts for both airborne and ground mobile forces in the electronics-intensive battlegrounds of the 1980s and 1990s. These concepts include the self-protect, low-probability-of-intercept aircraft, and a truly jam-proof, covert (spread-spectrum) communications network.

Enhanced capability data processors will allow systems designers to do more with the data from signal processing systems than display them for the human operator. The target detection, identification, classification, and tracking can potentially be performed automatically. The functional density may, in fact, allow the development of a fire-and-forget missile that can automatically adapt to the ever-changing battlefield environment.

However, this enhanced processing capability can do more than the quantitative "numbers crunching" of the past. The "computational plenty" of VLSI may offer the basis for new architectures capable of qualitative computation, characterized by the ability to perform the association, inference, and deduction operation of machine intelligence. In any case, the impact of VLSI on military systems will be measured by the quantity of added intelligence—through enhanced arithmetic or logical performance—that we can add to DoD systems.

Above all, the VLSI era can provide affordable, cost-effective weapons by minimizing both acquisition and logistics costs or by providing superior capabilities so that fewer high-cost systems are required to maintain adequate capabilities to respond to the ever-changing threats to our national defense.

II. THE DOD'S NEEDS: ELECTRONIC ISSUES

A. An Electronic Technologist's View of the Battlefield

The value of technology to the battlefield has been recognized since the bow and arrow was first used in combat against those with only the spear. Mankind's drive for self-preservation or self-defense drove the development of warfare technology as fast or faster than the development of the technology to address his other basic needs of food and shelter.

To fight a battle effectively, for either defensive or aggressive reasons, three elements are required. It is first necessary to have weapons as good as or superior to those of your adversary. Second, means must be available for moving troops and weaponry to and around the battlefield. Finally, there must be a means of communication, both to coordinate the troops during the battle and to maintain contact with those elements supplying support for the battle.

Over a very short period in recorded history, weapons have progressed from the spear to the gun to today's laser-guided missiles. The necessary mobility first provided by foot, a crude boat, or a draft animal has moved to the complex electronics-intensive ships, submarines, aircraft, and unmanned missiles of today. Communication remained the province of the foot soldier, or courier, until World War II. Although the telegraph was available during the Civil War, it was primarily used for communicating the results of the battle back to the home front—the stringing of wires in the battlefield prevented the effective use of the telegraph for communications among troops in battle. World War I was fought with much the same means of communication. However, the invention of the radio, the first real use of electronics in the battlefield, drastically changed the nature of battlefield communications. Communications were now more timely and available to all elements of the battle.

World War II brought many rapid advancements in electronic battlefield technology. Active radar and sonar were effectively used in warfare and great strides were made in communication and navigation electronics. The lessons of history are that technology is effective in warfare and that it is necessary to have a technological parity with your adversary or a technological edge if one is to survive the perils of warfare in the modern world. This effectiveness was confirmed in World War II, which was won as much in the factory as in the battlefield.

Since World War II the United States' defense posture has been based primarily on technological superiority, a key element of which is electronic technological superiority rather than the numerical superiority of the past. Even before Vietnam, the U.S. had turned to technology instead of troops and massive purchases of simpler arms for warfare, although neither the Korean War nor the Vietnam conflict demonstrated the potential of the electronics technology available. The political climate in the U.S. requires that we station hardware and electronics equipment for defense instead of troops. Economic and budgetary constraints and the American unwillingness to maintain a large standing army also force us to turn to technological solutions. The expectation is that advances in technology will lower the cost and/or advance the basic capability of weaponry for defense. This reliance on technology becomes particularly important when we consider that our major adversary, the USSR, is spending from $20 to $30 billion/yr more on arms than the U.S. and is expected to do so throughout the 1980s. Thus the thrust of electronics technology, in particular VLSI, on our national defense needs must address the political and economic realities of the world climate as well as meet the needs of the systems that provide the deterrent to adversaries through an effective tactical capability for fighting a war.

The United States' strategy for defense of the free world has two distinct operational scenarios—the strategic theater and the tactical theater. The Department of Defense's objective in the strategic theater is to deter any attack on the United States or its allies by maintaining a nuclear strike force that could devastate any adversary, even after suffering the damages of a first strike. A triad of forces—deep interdiction bombers, nuclear-missile-carrying submarines, and intercontinental ballistic missiles—is used to provide the nuclear war deterrent. These forces consist of over 500 long-range bombers (mainly B-52s) to carry nuclear weapons to the enemy, over 1000 nuclear missiles (Minuteman and Titans) de-

ployed in 8 states in the midwestern United States, and 41 nuclear submarines, each carrying 16 nuclear missiles with each missile having 10 war-
heads. Each of these strategic weapons systems is heavily dependent on
digital electronics for navigation, communication, guidance, and self-
protective surveillance and defense. In addition, a key element in our nuclear defense is the electronics-based early warning systems consisting of
both ground-based and space-based radar and infrared surveillance
systems. Digital integrated circuits provide for these systems plus the
necessary communication network to interface the national command
authority with both the surveillance systems and the nuclear weapon carriers or installations.

The advance in the strategic area that can be most impacted by the application of VLSI are fairly easy to identify: we need more accurate guidance systems to assure the effectiveness, or the accuracy on target, of nuclear missile responses; we need better surveillance systems with increased resolution, i.e., more rapid detection and identification of nuclear
threats with a zero false-alarm probability; we need enhanced self-
protection weaponry for the missile platforms to assure survivability for
an effective second strike capability; and we need more reliable and secure communication networks for command and control of these forces.

The communication network is one critical problem for the military that
enhanced VLSI electronics can vastly improve. The military has invested
over $10 billion in the World-Wide Military Command and Control
System, a computer network designed to keep the commander of American forces in touch with both strategic and tactical forces in the field.
Although the system's design was initiated during the Johnson adminis-
tration, it has yet to effectively pass final tests. Attempts to send messages during testing lead to abnormal terminations more than 60% of the
time. In addition, the capability of controlling and distributing vast
amounts of information over computer-based communication nets can
overwhelm the commanders who must read and interpret this information. Studies have shown that commanders would have to read as much as
800 words/min for 24 hr/day to keep up with the system. Clearly, added
intelligence directed to the sorting and interpretation of information is a
key element in our worldwide command network.

However, the cost of strategic forces is more dependent on the nonelec-
tronic elements of the platforms than on the electronic portions—the
VLSI impact will be in the integrity, that is, the fail-safe, reliable operation, of our advanced strategic forces.

The tactical segment of our national defense strategy—the Naval surface fleets and attack submarines, the tactical aircraft for air superiority,
the air-to-ground combat support and deep-strike interdiction, the tanks

and armored vehicles, and the troops, communications, and supply networks necessary to support these forces—is based on a defense plan that requires the capability to contain a major Soviet offensive into Western Europe, maintain control of the major world oceans, and still have the forces to support a limited policing or peace-keeping function elsewhere in the world. Prior to 1969, the DoD's planning was based on two major engagements—containment of both a Russian and a Chinese offensive plus a police action. However, the ideological split between China and the USSR and the political and economic lessons learned in the Vietnam experience reduced our national resolve to a more practical "war-and-a-half" principle. The very fact that the Soviet Union became a major nuclear power compelled the United States to plan for conflicts with conventional or tactical weapons.

The Soviet buildup of modern tactical forces, as illustrated by Soviet spending versus United States spending in Fig. 4, threatens worldwide political stability and the credibility of the United States to contain the USSR. Soviet commitment to extensive military production and development of key technologies is now paying off in large numerical advantages in key combatant weapon system capabilities, now rivaling American deployed units in operational effectiveness. As illustrated by Fig. 5, their numerical advantage is particularly threatening in the Western European theater; they have a 4 to 1 margin in tanks and are threatening air superiority with sophisticated electronics based air defense capability, and in the early 1980s they are expected to greatly improve their tactical aircraft-oriented air superiority weapons. The Soviet tactical capability for offensive electronic warfare and their heavy investment in survivable

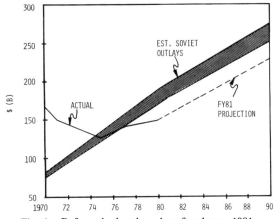

Fig. 4. Defense budget based on fiscal year 1981.

		US	USSR
MANPOWER	ACTIVE	2,000,000	4,000,000
	RESERVES	800,000	–
	CIVILIAN	1,000,000	–
NUCLEAR	BOMBERS	574	156
	SUBMARINES	41	62
	ICBM	1054	1398
TOTAL	WARHEADS	9200	5000
TACTICAL	AIRCRAFT	5026	5268
	SHIPS	175	230
	SUBS	81	234
	TANKS	10,000	43,000

Fig. 5. Comparative U.S./USSR military forces.

communications also seriously degrade the combat environment in which our future weapons systems must maintain their effectiveness.

In the maritime arena, the aggressive Soviet buildup in both strategic and tactical submarines and major surface combatant vehicles equipped with long-range tactical missiles increasingly threatens American sea control and maritime power projection. The momentum in the new construction of attack submarines is certainly with the Soviets, who currently have an estimated 3 to 1 numerical advantage over the United States. The technical environment for ocean control by American forces is being degraded by the continuing advances in submarine stealth and mobility afforded by the increased capability of USSR surface fleets for support. There is also a growing threat to our own antisubmarine warfare forces from the increased capabilities of the long-range Soviet Badger/Backfire maritime squadrons.

The tactical battlefield activity of the future can be expected to be intense, violent, and confusing. The frequency of engagements will be high but probably of short duration. Kill capabilities on both sides will be high—over 325 tanks were lost in one day of fighting in the 1973 Israel–Arab conflict. In Western Europe, military operations will be compounded by the presence of large numbers of civilians. Military commanders must be concerned with the control and protection of civilians as well as the control and movement of military forces and equipment.

In Western Europe, the density of military equipment will be immense. The Warsaw Pact planning doctrine calls for as many as 150 tanks in two 1-km-wide thrusts against a 7-km Western European front. A Soviet antitank defense front typically comprises a total of 26 to 30 antitank guns/km. NATO forces now average 5–15 major antitank guns/km and are trying to increase that number. There will also be a very high density of electromagnetic radiation from communications systems, radar, offensive and defensive systems, EW jammers, and others.

Western European forces, in particular NATO, are heavily dependent on electronics for command, control and communication, surveillance, target acquisition and identification, weapon guidance, and other elements for both air and ground systems. With few exceptions, NATO has an electronic capability superior to that of the USSR, in particular, in computers, infrared viewers and seekers, homing missiles, and automated fire power. Individual systems are also significantly better than the Soviets, but they are always acquired in smaller numbers. Future tactical parity with the USSR requires that we maintain the United States' technological lead in advanced electronics to achieve an increased weapons system effectiveness to offset the Soviet numerical advantages. In addition, we must do that with affordable weapons systems having reasonable life cycle costs in order to maintain this parity with smaller budget outlays than our adversary.

Thus there are five essential elements of the national defense posture that advanced electronics must address. The first is the need for a credible warning of an attack or impending attack through surveillance and intelligence capability. The second is the location and identification of enemy forces and the capabilities of those forces. The third is an adequate defense of the our attack, support, and supply forces. The fourth is the capability for a quick-reaction, high-lethality strike in depth on the battlefield. The fifth is effective command, control, and communications in a fast-paced, complex, confusing battlefield environment. These elements highlight the advances in the digital technology needed by the military systems. The increasingly complex and adverse operational environment of the 1980s and 1990s and DoD's fiscal, schedule, and management realities combine to create stringent performance, operability, availability, and affordability goals for future military systems. Electronics technology is of major importance in achieving the needed flexibility to meet a wide-ranging set of weapons system capabilities while also achieving a major reduction in military acquisition and operational costs.

B. Response to the Threat: System Capabilities

In the broadest sense, four mission objectives are set for new American weapons systems in the 1980s and beyond. These are: (1) superior surveillance, including enhanced capability to find and assess the targets in adverse battlefield and/or standoff environments; (2) high-lethality weapons, including precision targeting and terminal guidance of weapons to offset relatively massive enemy concentrations of force; (3) high mission sur-

vivability, including penetration, stealth, electronic counter countermeasures (ECCM), defensive measures, and weather and radiation immunity; and (4) effective management of a wide variety of information
through effective command, control, communications, and intelligence
systems. These objectives must be achieved in an environment requiring
very low life cycle costs relative to the complexity of the systems.

The technological challenges posed by the 1990 mission environment
are immense. In surveillance and target assessment, we must acquire and
sort high-resolution radar, forward looking infrared (FLIR), or passive
radio-frequency (rf) sensor data for target assessment at longer stand-off
ranges or with higher-speed weapons platforms. In target processing and
weapon delivery, we must incorporate multitarget recognition and tracking
algorithms for faster target assessment and handoff to smart weapons. We
must also decrease the overload of multitarget analysis on the operator.
Systems must operate in increasingly adverse environments. We must integrate jamming and countermeasure immunity into sensors, increase our
active rf stealth capabilities, provide very high data compression for
antijam linking of remote sensors, and increase weapons acquisition and
homing capabilities that resist countermeasures.

Improvements in systems availability and reliability are absolutely required. We must integrate multimission radar processing into smaller
packages to fit the diverse mission needs in small tactical aircraft. We
must also package wideband image processors for remote sensor applications and small portable battlefield units. We must integrate complex
target processing in small tactical seekers and attain very large MTBF
improvements in tactical processors in all systems. In addition, we must
enhance the effectiveness of weapon system operators by reducing operator load through alleviation of response requirements at critical decision
points, automation of threat analysis and prioritization, automation of
target location and characterization, and automation of operator training
to reduce the level of operator skill. Of course, we must also reduce both
the acquisition and the logistics costs of these increasingly complex
systems.

Although these extentions of systems capabilities are within the realm
of possibility for U.S. scientists and engineers working with VLSI, we
must not rely totally on technology without balancing the technology with
realism. During the Vietnam conflict, many U.S. aircraft attacking the
North were equipped with radar warning receivers that could detect the
presence of aircraft surveillance by the variety of defensive antiaircraft
radars employed by the enemy. However, the density of these emitters
was such that most systems were turned off by the pilots once the planes
were into North Vietnam—the pilots did not have the time to analyze the

complex display or audio indication of the threats. To be truly effective, these systems must sort the threats and offer minimal guidance to the pilot during the operation.

Technological capabilities do not always overcome larger numbers of less-sophisticated weapons. Tests of air combat capabilities for our newer air-superiority aircraft showed that, in one-on-one engagements against simpler aircraft, the new plane was clearly dominant. However, during tests of multiple aircraft against multiple aircraft, the results were more nearly equal. Most "kills" in these engagements were obtained without the use of the new complex equipment. Technology for technology's sake is clearly not the answer.

However, advances in VLSI will solve many of the weaknesses in existing weapons systems. As an example, consider the TOW (tube-launched, optically tracked, wire-guided) missile antitank weapon. A single soldier stands and fires the TOW and, by keeping his sight aligned on the tank during the missile time of flight, can assure a hit. However, the soldier is exposed to enemy fire during this time of flight. This exposure time might be eliminated by the proper application of VLSI technology using a FLIR sight and onboard guidance. The soldier can designate the target for the missile while retaining his cover and the onboard digital electronics and the FLIR detector could maintain the guidance toward the designated image.

Perhaps the greatest payoff of VLSI to the military will come from the deployment of more small and low-cost systems, which the U.S. can afford in larger volumes, instead of the larger and more complex systems. Perhaps VLSI can bring to the battlefield the same class of effectiveness that the single-chip calculator brought to the business and academic worlds. Examples are restricted only by one's imagination. Small, easy to deploy, "smart" mines that can detect the presence within its kill or damage range of a priority enemy target such as a tank or armored missile carrier or handheld spread spectrum data terminals for secure communication between troops on the battlefield are examples that come to mind.

In any event, automatic handling of huge volumes of information is the key to new DoD systems capabilities. In undersea surveillance, the end objective is to locate, identify, and track all submarines in an ocean basin. Acoustic signatures of individual submarines are known, and since the signals travel surprisingly long distances in an ocean, they can be collected but must be separated from the clutter of the ocean environment. Thus VLSI computers are required to handle the large number of signals and reduce the data to minimal necessary information.

Signal intelligence is another key area for VLSI application. In combat it is necessary to track, identify, and locate the emitters for the thousands of signals that exist in the battlefield. Here VLSI computers are required to sort these signals and prioritize threats for the battle commander.

Another application for VLSI on the battlefield involves a missile that could distinguish one object from another to allow the production of a truly fire-and-forget weapon. However, the data required to identify a particular target are immense and the correlation processing capability required on board such a missile approaches the processing capability of our current supercomputers.

However, the ultimate force multiplier attainable with VLSI may be an effective command, control, and communication system. Without such a system, all other systems and efforts focused on winning the battle may be ineffective. What good are complex tanks and aircraft with high-lethality weapons systems if they cannot be moved to the point of need? The features of such systems are well known. They must be jam-resistant and have a fairly low probability of message interception. They must be *fail-soft* so that loss of one piece of the network does not shut down the system. They must transmit data, both voice and digital, with low error rates and must have automated message-handling and data-reduction mechanisms. They should be fairly flexible to deploy and exist without wires or line-of-sight constraints. They must be highly mobile, easy to operate, and affordable. The technological understanding required to build such a system exists today and VLSI will provide the hardware technology to build the system and do so with confidence in the affordability and reliability that must exist in a deployed military system.

C. The Market for Military Electronics

Although the military portion of the worldwide semiconductor market eroded from the 30% share as late as 1970 to the 7–10% share in 1979, the military remains a significant market force. Suppliers to the DoD find the market attractive due to: (1) the long-term stability of the DoD's R&D and procurement actions; (2) the DoD's willingness to invest in the independent research and development efforts of those industries serving the market; and (3) the sheer volume of DoD electronic purchases.

The stability of the DoD marketplace has provided a much-welcomed buffer to many semiconductor companies during the recessionary periods that cause severe declines in the commercial and, more recently, the consumer market. The DoD's development cycles are long, but, once the system is released for production, a market base for the designed semiconductor component is assured for several years.

This stability also exists in the independent research and development funds that accrue to those industries serving the military through "cost-plus" contracts. The DoD takes much of the risks on these contracts but assures the contractor of funds to address the risks of other advanced research and development efforts. Some of the large commercial customers provide semiconductor companies' research and development funds, but these funds are usually dedicated to a particular customer need and are subject to the swings of the general economy. Semiconductor companies require a large, stable research and development budget to maintain the rate of technological advances achieved in the past 20 years.

Although the dollar value of the DoD's budget has remained constant in real terms (adjusted for inflation) during the seventies, except for the Vietnam years, the electronics portion and, particularly, the integrated circuits portion have continued to grow. In addition, political uncertainies in the world today have resulted in a DoD fiscal year 1981 forecast to increase U.S. defense spending throughout the 1980s at 4.4%/yr in real terms. This is up from a projected 2.5%/yr growth in DoD's fiscal year 1980 planning. The fiscal year projection is indicated in Fig. 4.

Much of the DoD's budget must, of course, be spent on the salaries and pensions for the 2 million active-duty military members and 800 million reservists as well as the one million civilians (Fig. 5). Almost 60% of the total budget now goes for these costs, as illustrated in Fig. 6. (Civilian pay is included in the 30% slice of the budget for operations and maintenance.) This leaves approximately 36% of the budget for weapons procurement (26%) and research, development, technology, and engineering (10%).

Some defense analysts maintain that this budget is much below that

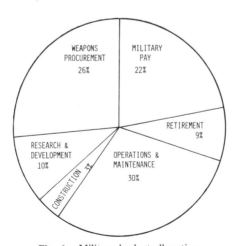

Fig. 6.   Military budget allocations.

necessary to implement the 1969 "war-and-a-half" strategy. The critics maintain that we are spending too little to offset the depreciation of our current military equipment by as much as 50%. The claim is that we need to increase procurement by 40%, operations and maintenance by 10%, and military pay by 20% to maintain our 1969 strategy.

On the other side of the ledger, there are analysts who maintain that our concepts for fighting a war are wrong, that massive, high-technology and high-complexity firepower aimed at destroying the enemy piece by piece cannot work. The claim is that our tactics must be those historically effective when larger forces are overcome by smaller forces: (1) disrupt the enemies' plans; (2) undermine the enemies' alliances; and (3) strike at the integrity of the enemies' system—never directly engage his armies or attack his cities. These critics want more reliance on simpler weapons and more attention to military maneuvers instead of military logistics. They also claim that we overestimate the effectiveness of Russian weapons just as we overestimate the value of some of our complex weapons systems. However, these critics do not wish to reduce the defense budget but merely to redirect the DoD's spending strategy.

One thing is certain. The United States is obtaining less and less tactical capability for more and more dollars. Although estimated inflation has been removed from our projection of DoD budget figures, inflation impacts more than just price. As prices rise, complex weapons systems procurement volumes are reduced due to fiscal constraints, thus decreasing production run quantities and, therefore, increasing unit price even more than inflation—due to the fixed overhead costs that must be paid over longer periods of time. This factor has increased the estimated costs for the total F-15 program by 90% since 1970, the costs of the F-16 by 100% in the same timeframe, and the costs of the A-10 program by a phenomenal 230% since 1973. In many cases the budget is used to keep the factories open instead of to deliver increased capability.

However, the electronics content of the research, development, technology, and engineering (RDT&E) and procurement portions of the DoD's budget continues to increase. More than 26% of procurement dollars and 34% of RDT&E dollars are electronics related. Figure 7 illustrates the estimated growth of the electronics content of DoD equipment by major system categories. The sums, although a small portion of the total DoD budget, are immense—rising from approximately $20 billion in 1979 to almost $30 billion in 1988, for a growth rate of 4.2%/yr (in FY81 dollars).

The electronics and communications segment will remain the largest factor in the budget, accounting for approximately $9 billion in 1988. However, its growth rate will be a fairly uniform 4.6%/yr. In com-

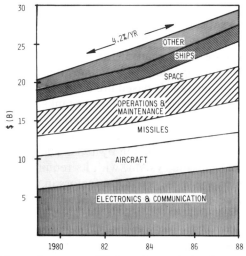

Fig. 7.   Electronic content of DoD equipment in fiscal year 1981 dollars.

munications, the U.S. must provide the capabilities for reliable, *fail-soft*, and covert data and voice networks, as indicated in a previous section. These systems must allow for an update of equipment at all levels within our combat forces—from the foot soldier and battlefield vehicles to the complex tactical command centers surrounding the battlefield. Antijam communication links are the key element here, as is the computerized capability for data reduction. Antisubmarine warfare, with its huge appetite for signal processing, will be a major activity in this and other market segments as will signal intelligence.

The largest growth will be in the space and the ordnance and vehicles segments, each of which will more than double over the 10-yr period (10.7%/yr and 9.4%/yr, respectively). Ordnance and vehicles will remain a small portion of the total (approximately 3.5%), but space will account for more than 10% of the total, or approximately $3.5 billion, by 1988.

The activity in the space segment comes from the need for enhanced surveillance, communications, and navigation for military systems. Major development programs are ongoing in satellite communications (SATCOM), navigation (GPS, or Global Positioning System), and surveillance, with both infrared (Project HALO) and radar (SPACED-BASED RADAR) sensors. VLSI is extremely important to these programs since space systems depend heavily on affordable, reliable electronics with system antijam and self-healing properties. By 1988, more than 60% of the DoD's space budget will be for electronics.

Although smaller, the ordnance and vehicle segment of the budget remains attractive to electronics suppliers. The electronic content of combat vehicles is expected to grow from 25% in 1980 to more than 40% by 1988 in order to increase the firepower per vehicle of our forces. The DoD plans to procure more than 350 of the new electronics-intensive XM-1 tanks, and major upgrades of existing M-60 tanks through the installation of more automated IR-based fire control systems are required to overcome the Soviet numerical advantage in tanks.

The missile segment will also be a major market through the 1980s. The growth rate of this segment is expected to be about 5.7% annually, reaching a total market of $4.4 billion by 1988. The development and deployment of the MX missile system, with its requirements for enhanced guidance and control systems, are major factors in this segment. In addition, the strategic defensive systems are due for upgrading with new technology to assure effective protection of our nuclear responses.

Air-to-air missiles will continue to receive high levels of funding. Air superiority will remain a key concept with DoD defense planners, as the performance of air-to-air missiles in Vietnam brought up many questions. In theory, the probability-of-kill ratio was 95% for these missiles; in practice only 8% managed to hit their target. New techniques for more realizable accuracy at greater stand-off ranges are required in this market.

Surface-to-air defense missiles will continue to gain funding. The effectiveness of these weapons was demonstrated in the 1973 Israel–Arab conflict, first by a high kill ratio on attack aircraft and then, as the aircraft pilot and mission planners became more wary of its effectiveness, by reducing the rate of air attacks.

Another growth area is in defense suppression missiles. Effective missiles that home in on radar emitters have been in the DoD's inventory for more than 15 yr (i.e., the Shrike). New defense suppression missiles that can be fired at greater stand-off ranges, prioritize and select one target out of many, and home in on the selected target in a vulnerable area, countering any electronic countermeasure, are much sought after. These missiles will have combinations of IR, active radar, or passive rf sensors.

Also in the missile segment, we will see a frenzy of activity to develop an effective antitank weapon. Techniques that utilize advanced IR or millimeter-wave radar sensors and are dependent on a high throughput electronics capability to perform image processing and target recognition functions are obtaining increased DoD development funding.

The electronics content of the ship segment will be one of the smaller growth segments primarily due to the fact that our defense posture is based on few, but complex, surface and subsurface platforms. The Trident submarines cost approximately $1.6 billion apiece and the nuclear carriers average $2.2 billion each. Obviously, we cannot afford large volumes of such platforms. However, large sums will be spent on defensive systems for these platforms, since the loss of just one aircraft carrier or Trident-class submarine is a major blow to our naval forces in both expense and firepower effectiveness. Remember, our naval posture is based on only 12 carriers and 12 Trident-class submarines.

The aircraft segment is expected to remain flat over the decade. Even R&D activity in aircraft-related systems will drop in the early 1980s but pick up again before the 1990s. No new aircraft starts are currently envisioned, but defense problems remain that must be addressed, perhaps by new or modified aircraft. Although the B-1 deep-strike bomber development was halted in favor of extending the life of the B-52, support for this development or a derivative may regain momentum due to the worsening state of world affairs and the necessity of providing a platform for the new air-launched cruise missiles (ALCMs). A $4.4 billion program for production of these missiles has just been awarded.

There also exists some pressure for a new transport aircraft, designated the C-X, to assure the DoD the mobility to deliver a quick-reaction, effective force to any global trouble spot in a matter of hours. Disappointments with our C-5 aircraft and the limited capacity of the C-140 class of carriers are behind the move toward a "new" airframe.

However, in light of the budgetary constraints and the expense of new aircraft (the cost of aircraft has increased by two orders of magnitude since World War II), these needs will more than likely be met by upgrades to existing platforms throughout the 1980s. Conversion in lieu of procurement (CILOP) will be the buzzword of the 1980s in the aircraft market. As we lengthen the lifetime of aircraft, technology insertion, primarily increased-capability electronics, will be the key to meeting the system requirements.

The B-52 aircraft will certainly be upgraded in its electronics capability to assure its survivability in performance of its strategic role. The U.S. attack aircraft must be upgraded for all-weather, day–night, low-level penetration capabilities to survive and be effective in the battlefields of the future. Enhanced terrain-following capability is of necessity here: the aircraft that survives will probably be the one that flies in near ground zero, pops up to deliver its weapons, and returns to safe airspace again near ground zero. Passive imaging techniques offered by IR and passive rf systems will be keys for ground support attack fighters, while low probability of intercept radars will be required for survivable air-superiority aircraft. In the ocean-control aircraft segment, the range and area of coverage of the U.S. squadrons of S-3 and P-3 aircraft must be extended through greater electronic capabilities at even lower cost since new aircraft can do little to aid this coverage problem. Thus electronics will have a major share of the total expenditures for aircraft in the 1980s.

The near-term elements of the RDT&E portion of the DoD's budget, which provide for the science and technology base for the DoD's needs, have been precisely identified by Dr. William J. Perry, undersecretary of defense research and engineering. His FY81 posture statement said the objective of DoD's science and technology program is to "maintain a level of technology supremacy which enables the United States to develop, acquire, and maintain military capabilities needed for national security."

Specific goals for the efforts are also indicated. The first is to provide for real growth in the technology base by providing the infrastructure so necessary to the steady evolutionary growth of military capabilities. Our technology growth must be both comprehensive and diversified and is, therefore, very dependent on stable, dedicated R&D funding. The second goal is to exploit the use of advanced technology developments to provide for more rapid and effective transition into military systems. This is to be accomplished by demonstrating the technology base in useful applications, thereby illustrating its effectiveness or determining new or modified applications to exploit its capabilities. The thrid goal is to expedite the technologies that are of prime importance to maintaining our technological lead or that exhibit the potential for high military payoff. The technologies to be emphasized in the 1980s will be heavily electronics-oriented, as indicated by the following identified efforts.

The first effort is in the development of precision-guided munitions (PGM). The effort will concentrate on the sensors and alogrithms and digital signal processing required to achieve an autonomous, adverse-weather capability for accurate, low-cost guidance systems to deliver the terminal seeker to the point where it can acquire and home in on the target.

The second is the very high speed integrated circuits (VHSIC) program begun in 1979. This program, to be described in a later section, is a six-year, $210 million effort to achieve a 100–200-fold improvement in integrated circuit technology and apply the technology for maximum benefit to the Department of Defense.

The third thrust is the development of directed-energy capabilities such as high-energy lasers and particle-beam systems to determine the potential payoff of such revolutionary warfare technologies.

The fourth effort is the embedded computer software technology initiative. The increased performance and reduced costs associated with computers have greatly multiplied the uses of computers in military systems.

This program will concentrate on the associated software advances that are required to maximize the effectiveness of computers in military systems. The intent is to develop the automated software technology necessary to improve the responsiveness, reliability, performance, and cost characteristics of military systems throughout their total life cycle.

Manufacturing methods and materials technology funds will also be used to address electronic requirements and advances. The funding for the science and technology base is expected to show a real growth of more than 6% in the early 1980s.

Of course, the value of semiconductors in military systems is a small portion of the electronic equipment costs. As illustrated in Fig. 8, the semiconductor content of electronic equipment has risen from less than 2% in 1970 to 3% in 1980 but is expected to increase to 5% by 1990. Although the percentage will rise faster than it has in the past, its growth will be less than the average for all segments of electronic equipment and will be much less than the 12% content in the computer market and the expected 9% content in the consumer market.

However, the military semiconductor market is expected to cross the $1 billion level in 1980 with the integrated circuits elements alone crossing the $1 billion level by 1982. The growth rate of the military integrated circuit market is expected to be as much as 18.5% during the 1980s, as indicated in Fig. 9, and may reach a $4 billion annual market by 1990. At this point integrated circuits will account for more than 80% of the semiconductors sold to the military.

The breakdown by product types in the military semiconductor market is graphically depicted in Fig. 10 for the years 1980, 1985, and 1990. The percentages for the integrated circuit portion of the market are shown.

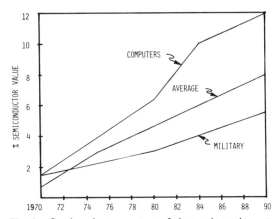

Fig. 8. Semiconductor content of electronic equipment.

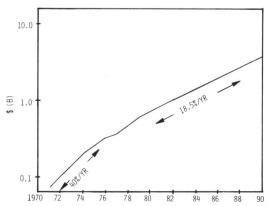

Fig. 9. Military integrated circuits market.

The remainder of the semiconductor products for the military market consists of power devices, optoelectronic devices, and small signal transistors, and although they will account for 30% of the market in 1980, this share will be reduced to 13.5% by 1990.

As is the case in the commercial marketplace, microprocessors are becoming more commonplace in end-user equipment, and as the content of programmable processors in systems grows, the use of memory products will rise dramatically. During the late sixties and early seventies, as the semiconductor industries were making great strides in reducing the cost of, first, bipolar logic and, then, MOS logic, the ratio of memory cells to logic gates in end-use equipment fell dramatically. However, with the advent of solid-state memories in the mid-seventies and large-scale microprogrammable logic functions in the late seventies, the ratio once again turned upward. In 1980 the cost per bit of memories, particularly ROM memory, was very attractive compared to the cost per gate of logic, largely due to the universality of memory components and thus a sharper learning curve. Thus the trend is toward higher memory-cell-to-logic-gate ratios in systems. As depicted in Fig. 11, the ratio fell from 40% memory-to-logic in 1965 to a low of less than 5% in 1975 but is expected to return to 40% or more by 1990.

The military marketplace will follow this trend, with memory components accounting for more than 45% of the total military semiconductor market by 1990. Of this memory, it is expected that 60% will be MOS technology and 40% will be bipolar. This ratio could lean even more toward MOS if military programs continue to waive the temperature and radiation requirements in military systems to obtain the benefits of the higher-density MOS memories. Bipolar memory in the past had a strong commercial market due to its higher-speed characteristics. However, as

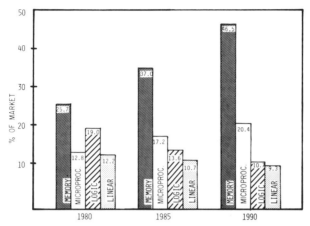

Fig. 10.   Percentage of total military semiconductor market by product.

we continue to shrink technology to the 1-μm barrier, the speed (access time) differential between bipolar and MOS will rapidly disappear, as will the difference in reliability.

The same move toward MOS may also occur in the microprocessor market segment due more to the weight of MOS development activity versus bipolar than to the learning curve factors. However, the current forecast is for bipolar technologies to retain an approximate 80% share of the microprocessor market. The market segment itself will rise from $130 million, or 12.8% of the total military semiconductor market, in 1980 to more than $1 billion, or 20.4% market share, in 1990.

Logic will continue to be dominated by bipolar technologies, with some

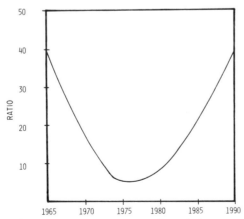

Fig. 11.   Ratio of memory cells to logic gates.

Dennis R. Best

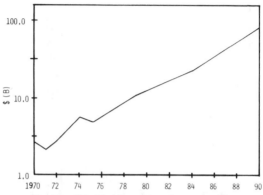

Fig. 12. Worldwide semiconductor market.

inroads by CMOS, due to the static nature of these technologies. However, logic will reduce its percentage of the market from 19% in 1980 to 10.7% in 1990 as we learn to apply microprocessors to the military marketplace and microprogrammable signal processors become available with the advent of VLSI. Much of this logic, in particular, that which is able to take advantage of VLSI integration levels, will be applied as logic or gate arrays which are user-defined in ways similar to the user-programming of microprocessors today.

Linear integrated circuits will remain a stable market, growing at 13.5%/yr over the decade. However, they will lose slightly in percentage of the market, going from 12.3% in 1980 to 9.3% in 1990. The effective transition of high-speed gallium arsenide integrated circuits from laboratory to production might significantly alter this projection.

However, a glance at the dollar volume of the total worldwide semiconductor market, shown in Fig. 12, illustrates a hard fact concerning the military semiconductor market. Its share of the total semiconductor market is continuing to fall from 8% in 1980 to approximately 5–7% in 1990, but a $5 billion annual market for semiconductor products will be serviced.

III. VHSIC—ROADMAP FOR SUCCESS

A. The DoD's VHSIC Program—A New Approach to Military Electronics

Over the past decade the Department of Defense has followed a deliberate policy of not giving heavy funding to integrated circuit programs. The DoD adopted this policy in the belief that commercial market forces would yield the integrated circuit advances necessary to military systems. However, as it became clear that LSI technology was not rapidly translating into advances into military equipment, the DoD reversed its policy and is now planning a multiyear multimillion dollar investment to expedite the advance of integrated circuits from the LSI to the VLSI era.

The initiation of the VHSIC program was born out of the DoD's current judgment that the industry is not presently oriented toward meeting the DoD's current and future needs. The industry trend is directed toward high-volume commercial applications in which increasing integration is used for increasing functionality as opposed to increased performance. These high volumes allowed commercial applications to use custom LSI solutions since the high initial design costs were written off over many production units. The military tends to buy small volumes of a diverse set of custom integrated circuits and, certainly, the DoD market has many specific applications requiring the ultimate in integrated circuit performance.

In addition, the Department of Defense felt that industry is not sufficiently oriented toward qualifying integrated circuits for military applications or toward developing integrated circuits with clock speeds sufficiently high to perform the current and projected real-time, high-speed, signal and image processing functions. Also, in contrast to commercial requirements, the military depends on more built-in test-and-fault tolerance techniques in order to provide acceptable logistics expenses as well as more fail-safe capability for unique electronic applications. At present, new approaches to these uniquely military problems are not being pursued sufficiently to minimize the need for customization in military systems.

The VHSIC program differs significantly from commercial VLSIC efforts. The VHSIC program is concentrating on the development of ICs for broad classes of military systems, particularly those for which there is no comparable commercial or industrial need. It is also placing a major emphasis on achieving new system and component architectural concepts, which will minimize the need for design customization and hence reduce cost in both acquisition and logistics. There will also be a major emphasis on increasing the real-time system throughput of integrated circuits, which will require not only the higher chip complexity associated with increased integration but also the higher clock rates to achieve this capability: thus the name very high speed integrated circuits. In addition, there will be a major emphasis on military environmental requirements such as performance over a high temperature range, performance in expected battlefield radiation requirements, and very high reliability. There will also be exploitation of the higher chip complexity to achieve the necessary built-in test and fault tolerances to improve the basic reliability and availability of military hardware. The VHSIC program will balance out the industry's trend toward VLSI and concentrate on those technology tasks essential to achieving military goals.

The VHSIC program will extend over a six-year period and the funding will average approximately $36 million/yr for a total of about $210 million. As illustrated in Fig. 13, the program is divided into four parts—Phases 0, I, II, and III. Phase 0 is the program definition part of VHSIC. It will consist of analyses of approaches toward meeting the goals of Phase I and Phase II programs. It will include, for example, system and subsystem analysis, definition of integrated circuit functionality, definition of the military technologies, definition of computer-aided design and test requirements, and such experimental fabrication and testing of devices, layouts, and processing techniques as necessary to form a basis on which to make follow-up decisions. The output of this phase will be a definitive plan and approach for the follow-up phases.

Phase I is conceptually divided into two parallel efforts: Phase Ia and Phase Ib. Within about four years after the start of the program, Phase Ia will result in the construction of complete electronic brassboard subsystems, using integrated circuits developed to 1.25-μm minimum feature sizes. The goal for this technology is to obtain an equivalent gate–clock frequency product exceeding 5×10^{11} gate–Hz/cm^2. This figure assumes a maximum power dissipation of 3 W/cm^2 and a minimum clock rate of 2.5×10^7 Hz. Phase I will also result in the establishment of a pilot line production capability for devices with these feature sizes and will demonstrate the applicability of the design tools, simulation aids, architectural concepts, testing techniques, and packaging developed for VHSIC technology.

Phase Ib, running concurrently with Phase Ia, will consist of initial efforts to extend the state of the art of IC fabrication to submicrometer feature sizes and the associated higher performance and gate densities. It will address the problems of submicron technology and identify promising approaches to solve the problems.

Phase II is also conceptually divided into two parallel parts. Phase IIa will provide system demonstrations using the electronic brassboards developed in Phase Ia to demonstrate the capability of the 1.25-μm integrated circuits. Phase IIb will provide a capability for further improvements in system performance by projecting subsystem design concepts to higher performance levels by extending the state of the art of IC technology into the submicrometer feature sizes. The end goal for IC fabrication is to reach 0.5-μm feature sizes. The equivalent gate–clock frequency product goal will be increased to 10^{13} gate–Hertz/cm^2. Phase IIb is essen-

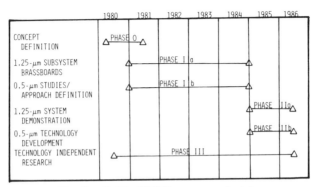

Fig. 13. DoD's VHSIC program schedule.

tial to find problems encountered in crossing the 1-μm barrier established by the use of conventional optical lithography. Additionally, this phase is considered essential to ensure that the DoD remains at the leading edge of technology in order to meet the most advanced and presently projected needs.

Phase III is a six-year program, running in parallel with Phases 0, I, and II, consisting of efforts that support or supplement the other phases and provide new or alternative directions not specifically included in those other phases. Phase III is considered essential to provide needed flexibility in the program to stimulate additional innovation and to incorporate the broadest possible participation of industry with the program. Phase III shall consist of many shorter programs of limited scope and shall concentrate on key technologies, equipment, or tools. Examples include the development of basic IC fabrication technologies such as e-beam and/or x-ray lithography and associated resists and resist-removal equipment. Other Phase III efforts may include advanced packaging concepts, advanced automatic testing equipment development, improved semiconductor materials, establishment of design standards and interfaces, and development of innovative architectural approaches.

Systems analysis and eventual systems demonstrations are considered a key part of the VHSIC program. Demonstration of advanced ICs will expedite the insertion of this advanced capability into future military systems in a timely and affordable manner. It will also provide tangible evidence of the value of the ICs to the DoD system community and will allow the military to realize a near-term return on the $210 million DoD investment. Also encouraged is the establishment of a military market for the integrated circuits, which is necessary to maintain and improve the manufacturing of complex circuits.

The high-priority subsystems to be analyzed and brassboarded during the program will be identified in the early stages by the three services. These will be selected to demonstrate that the VHSICs clearly meet the high-throughput signal and data processing needs of the services as applied in such systems as satellite, Cruise missiles, advanced radar, command, and control systems, wideband data communications, antisubmarine warfare, electronic warfare, and intelligence. The systems will be selected to allow demonstration of military-related functions previously precluded by computational limitations or platform environmental constraints and will also allow demonstrated reduction of life cycle costs designed for the military and typical DoD environments, such as development, acquisition, operation, and maintenance expenses.

To ensure that the integrated circuits will have a broad applicability to the military, VHSIC contractors will be expected to analyze at least three unique subsystems. The subsystems are to be selected such that at least one major need of each of the Army, Air Force, and Navy is considered. In addition, to achieve a maximum degree of commonality in the IC design, the three subsystems must have diverse applications; for example, radar for one service, communications for another, and electronic warfare for the third. By emphasizing the systems over the entire life of the program, the DoD hopes to assure a technology that solves the real-world problems.

B. The DoD's VHSIC Program—Expected Payoff

The payoff of the VHSIC program is expected to be far more than just technological advancement. The DoD has stated that the four main reasons for initiation of the VHSIC program are as follows: (1) to introduce advanced integrated circuits into future military systems in a timely and affordable manner, minimizing customization requirements; (2) to increase America's lead in military ICs over our adversaries by several years; (3) to provide highly advanced military capabilities; and (4) to provide a high return on investment (ROI) in life cycle cost.

The first reason addresses VLSI availability to the military. Recognizing the military's loss-of-market share to the fast-growing commercial market and thus the smaller profit motive for the IC manufacturers, the DoD hopes that the VHSIC program investment will spur commodity IC suppliers to address the military's unique performance and environmental requirements. The expection is that this infusion of funds will allow the semiconductor industry to address the complexity and customization problems that have prevented widespread use of LSI in military systems. The DoD expects that 75% of the VHSIC program will provide either direct or indirect fallout to the commercial marketplace. This should help catalyze the VLSI advances by providing for feasibility demonstrations of the advanced ICs and by establishing an early viable market.

The $30 million budgeted annually for the program, although representing less than 10% of the industry's annual expenditures on semiconductor R&D, is expected to give the semiconductor manufacturers the incentive to pull forward to the 1-μm fabrication barrier by at least two to three years. There exists in the United States a three-year lag between the introduction of a commercial IC and its use in military systems. Such a lag reduces our technological lead in electronics over the USSR. There is evidence that the Soviet Union is making rapid gains in integrated circuit technology for the military. In the USSR, military development is given preferential treatment over any electronic development for a commercial use. Since the DoD's defense posture is based primary on technological superiority, the DoD considers this shrinking of our technological lead unacceptable and one that must be reversed.

The third reason for the program—advanced capabilities—indicates the DoD's recognition of the importance of integrated circuits as a force multiplier in military systems. The demands on processing capability in such diverse military fields as radar, missiles, communications, undersea surveillance, and satellites continue to grow. Figure 14 illustrates the performance improvements in electronics-based military systems that must be achieved to respond to the threats estimated to exist by 1990. Note that undersea surveillance, the end goal of which is to locate and track all submarines in an ocean area, would require processing capabilities that are greater than twenty times the capability of existing supercomputers.

The desire for new operational capabilities in systems will also demand increased computational capability. Current air-to-ground or air-to-air missile systems require positive radar or visual contact with the target, thus placing the weapons platform in jeopardy since we must assume that

APPLICATION	PROCESSING IMPROVEMENT	VOLUME X POWER REDUCTION	RELIABILITY IMPROVEMENT
SIGNAL INTELLIGENCE	100X	3X	>10X
RADAR	50 TO 100X	4X - 10X	2X - 10X
WEAPONS TARGETING	100X	16X	10X
IMAGE PROCESSING	200X - 500X	-	-
WIDEBAND COMMUNICATIONS	50X - 70X	4X	100X
ASW - GLOBAL SEARCH	4000X	-	-
ELECTRONIC WARFARE	1000X	-	-

Fig. 14.   Projected DoD's system performance improvement needs.

our adversaries have nearly the same capability. If we could achieve the performance in a missile seeker required to distinguish priority targets such as tanks from clutter or nonpriority targets such as trucks, then we could field a truly fire-and-forget missile which could be launched with little risk to the weapons platform. The ability to do this is estimated to require the performance of about 100 million instructions/sec, or more than 300 times today's best single-chip microprocessor.

Consider the digital performance required to utilize the image generated by the infrared detectors. The movement of FLIR detectors into high-volume production has created a new buzz word in the industry, *image processing*. (Texas Instruments alone produced more FLIR detectors in 1980 than existed in the world prior to 1980.) If we assume that we wish to display the image on a TV screen of 360 lines × 480 pixels/line and wish to update the image at a 60-Hz rate while performing 10 operations/pixel to effect a neighborhood operator image enhancement algorithm, then the processing required is 360 × 480 × 60 × 10, or 103.68 million operations/sec.

The fourth reason for the program, which is probably the most important to our national defense, is the return on investment to the military, or the cost payoff. The military, particularly in light of the current fiscal dilemmas, wishes to obtain the cost benefits of large-scale integration that have accrued to the commercial users of LSI equipment. The DoD must place increased emphasis on life cycle costs since military systems are often very complex and system downtime during periods of need is unacceptable to our national defense. Also, the costs of ownership, or logistics and maintenance costs, are often much greater than acquisition cost, or development and procurement costs, for complex, long-life military equipment.

There are many ways increased levels of integration can save life cycle costs. Consider a typical digital missile guidance system. Almost half of the electronic costs of such systems is in memory devices and most of the other half is in printed wiring boards (PWBs) and their connectors. If the number of components is reduced by a factor of two, the number of PWBs and connectors and assembly operations is reduced such that production costs would reduce by at least a factor of two even if the component costs remain the same. In addition, since system reliability is a function of the number of components and is extremely sensitive to the number of connectors, since these tend to be the most unreliable elements in systems, the system reliability and thus the cost of maintenance would improve by a factor of more than four. The cost of spares, typically 10–30% of system cost over the life of a system, would also be reduced.

Another element of cost savings is the volume or weight savings afforded by VLSI. Over the typical 20-yr life of military equipment, a pound of weight in the electronics will cost $2000 in an aircraft and $10,000 in a satellite. Thus, a 10-lb weight savings in the F-16 aircraft electronics bay will translate into $24 million.

However, on the F-16, space is more of a problem than weight. If we use VLSI to increase the survivability of the aircraft through an enhanced-attack radar capability, then the expected attrition rates in battle would be reduced. If the attrition of the aircraft on a particular mission is 25%/1000 sorties before increasing the capability and then we reduce the attrition to 10%, if we assume that we have 100 aircraft performing 2 sorties/day in a 5-day conflict, then we would lose 25 aircraft of present capability and only 10 with advanced capability. Since the replacement cost for the aircraft and trained pilot is approaching $18–$20 million, a 5-day savings of up to $300 million would accrue. The decreased size and weight of VLSI can also provide smaller missiles of increased range, thus enhancing the effectiveness of the missile and allowing fewer weapons platforms (aircraft) to carry larger payloads (missiles) on each mission.

There are many other ways to use VLSI to reduce costs. Reduced power levels save $20/W in aircraft, $2000/W in satellites, and $3000/W in field-portable, battery-powered equipment over a typical 20-yr life cycle. Use of silicon areas for self-diagnosis can save millions of dollars by reducing maintenance costs and increasing weapons system availability.

However, the greatest cost savings will accrue only if we use this increased silicon area to manage system complexity in all the areas of the life cycle cost formula: in development, by allowing the systems designer to design with high-level functions instead of NAND gates and flip-flops; in production, by providing components that are readily available with short delivery times from SC suppliers, are amenable to automatic assembly techniques, and are designed to aid the test and integration portions of the manufacturing cycle; and in deployment, by providing higher reliability, self-diagnosis or maintenance aids, and, most importantly, increased system capabilities or force multiplication.

IV. THE VLSI ENVIRONMENT

A. The Technology

The technologies of the 1-μm VLSI era will be those, or derivatives of those, we are familiar with today. The emergence and acceptance of new technologies are assumed due to the weight of R&D effort and funding being applied to existing technologies. However, the competitive pressures of the semiconductor marketplace will search out and determine these 1-μm technologies.

Since the military is a small percentage of the total IC market, the successful supplier of military-grade ICs will find a way to meet military requirements that by adaptation from the commercial mainstream technologies. The military simply cannot affort a military-only technology.

Of course, NMOS is receiving the most attention today, largely due to its cost and density advantages in the high-growth microprocessor and memory markets. It has demonstrated circuit speeds as good as any other high-density technology and, as processing approaches 1-μm dimensions, is expected to attain the speeds of the bipolar devices. Comparison of present static NMOS memory access times with those of their lower-density bipolar counterparts illustrates this fact. However, there are problems to be solved as we shrink NMOS. The limited drive capability of all MOSFET technologies, which degrades circuit speed and increases layout complexity, and the very thin gate oxides may cause severe design, manufacturing (defects and process control), and performance (speed, reliability, and temperature and radiation tolerance) problems.

For these reasons, there has been considerable study of submicron MESFET devices. The MESFET devices, which replace the gate oxide of MOSFETS with a Schottky gate, overcome some of the NMOS problems and may be a better submicron technology.

In any event, NMOS will probably remain the dominant memory technology, even in the military market. Actually, NMOS memory reliability has increased as memory densities have increased, and we in the military equipment business have discovered that lower junction temperatures mean higher system reliability and have started designing to keep ambient temperatures below 85°C. The only problem may be NMOS radiation performance at submicron levels. Circuit design and materials techniques may overcome this problem, but at the cost of deviation from the main stream.

Currently CMOS is receiving much more attention than it has in the past, primarily due to its low power and static design characteristics. The preponderance of activity is in the consumer market, but recent improvements in speed and breakdown voltage have made it a force in the telecommunications and military markets. The progress in silicon gate CMOS has effectively decreased the efforts in CMOS/SOS, since it is approaching CMOS/SOS capabilities with fewer of the manufacturing and materials problems. In the 1-μm era, CMOS will receive even more attention since the circuit density penalty (CMOS at 5 μm has one-half or less the density of NMOS) will be alleviated as will all the processing complexity. The size of the gate is less important at 1 μm since interconnection of the gates consumes more of the chip area than the gates themselves even at 5 μm. Also, the static properties of CMOS allow it to serve the logic array market. However, CMOS is still a MOSFET technology and may exhibit severe temperature and radiation problems at submicron dimensions.

Bipolar, in particular, high-density bipolar, will continue to be a major market force in the VLSI era. Integrated injection logic was the first of the low-power, high-density, bipolar technologies and currently provides the only JAN-qualified, military-temperature, radiation-hard, 16-bit microprocessor on the market—the Texas Instruments' SBP9900A. A survey of the literature indicates that advanced, high-density bipolar technology has been demonstrated as complete 1-μm, minimum-geometry LSI circuits by more than one semiconductor supplier. However, the VLSI bipolar technology will probably be a derivative of I^2L such as Schottky transistor logic (STL) or integrated Schottky logic (ISL). Although not as dense as I^2L at 5 μm, these circuits approach ECL speeds (2–3 nsec/gate) with a 500 × advantage in speed–power product over ECL. Again, as gate sizes become less of a factor on slice areas, these technologies should become the mainstream VLSI technologies for processor and gate array logic in both the computer and the military markets due to their performance, temperature, and radiation advantages.

The other chapters in the volume will cover in more detail the exact properties and problems of technologies and manufacturing techniques as we approach submicron dimensions, and there are formidable problems. In semiconductor materials, we shall face limitations in purity, uniformity, and surface quality. In lithography, we shall be constrained by the wavelength of light and severe alignment problems. Optical projection and direct step-on-wafer (DSW) techniques will carry us to the 1-μm limit, but to obtain submicron geometries, we will have to use direct slice writing e-beam techniques. This, in turn, will require innovative solutions to overcome the throughput (wafers per hour) limitations of e-beam equipment and the development of new, higher-speed resists for e-beam patterning. Imperfections that can currently be ignored may become critical at submicron dimensions. Current metal-to-silicon contact techniques may become completely unacceptable as we increase interconnect levels to handle the densities attainable with submicron technology.

The design of circuits with submicron dimensions also poses formidable problems. Empirical modeling techniques may become unacceptable for this class of circuit. Simulation techniques to add process-parameter effects on electrical parameters must be developed. The complete circuit must be modeled from functional requirements and compared to the layout in each step of the design process. New techniques for modeling complex circuits must be developed, since modeling a relatively simple

1000-gate processor with conventional truth table techniques requires more that 2^{100} bits of information, and an exhaustive pattern test of such a model or the resulting circuit would be prohibitive on the largest and fastest of today's computers.

These problems will be solved, but two effects of these problems will have significant impact on the military. First, the investment required to successfully design and produce the VLSI class of component may restrict the number of American manufacturers that can deliver this capability to as few as three companies. Wafer fabrication facilities alone, which cost $10–$15 million today, will increase to $30 to $40 million in the VLSI era. Second, because of the high cost of developing and using the software-intensive computer-aided design, manufacture, and test systems, custom IC solutions that are used today to solve unique military system problems may be prohibitive in the future.

B. The Components

Due to the very high design, manufacturing, and test costs associated with VLSI, the military systems designer will rely on a small set of programmable system components (PSCs) for the bulk of his electronic component requirements. These components will include microcomputers with much higher performance and greater functional capabilities than today's versions, microprogrammable signal processors similar to but more powerful than today's precursor, the Intel 2920 for the more-structured, higher-throughput vector processing in military systems, logic arrays (or gate arrays) that are programmable through the functional definition of their interconnect structure, and the necessary memory components, i.e., RAM, ROM, and EPROM, to provide for complete designs.

The component capabilities achieved within the next decade will be a function of the trade-offs made in the use of the inherent performance of submicron devices. Increased functional density may be obtained by sacrificing gate speed, or functionality may be reduced in order to provide a higher yield device size. However, we expect to see a full 32-bit microcomputer with up to 1 million bits of on-chip RAM and ROM within the decade. Single-chip microcomputers with ten times the throughput of current devices, or millions of instructions per second, and the memories required to support these rates will be available. Micro-signal processors having 10–20 times the microcomputer throughput will also be available.

There are those who maintain that the semiconductor industry will develop standard design rules and provide computer-aided design systems to allow every logic designer to become an IC designer and create an efficient custom IC for his particular system requirement. I disagree with both the potential and the principle of this approach. The potential is weak since this would, in effect, require a freeze on technology and would preclude much of the learning curve benefit that comes from continued high-volume production of the same designs. Also, the continued heavy investment in computer-aided design, test, and manufacturing (CAD, CAT, CAM) has reduced cost per gate of IC development by two orders of magnitude since the 1960s but the number of gates per IC has increased 1.5 times faster than this reduction. The principle is wrong since the ultimate value of VLSI is in increased people-and-asset effectiveness not increased circuit-to-application optimization. As we increase the complexity of systems, with more gates and memory bits, we cannot afford to design them on a gate-by-gate basis. To increase productivity we must design systems at an ever-higher level. In the sixties, I designed with transistors, resistors, and capacitors; in the early seventies, I designed with NAND gates and FLIPFLOPS; in the late seventies, I designed with register files, arithmetic and logic units, and microcontrollers; in the early eighties, I want to design by selecting complete algorithms that have previously been or are currently being coded for programmable system components; and by the late eighties, I hope to design by specifying subsystem functions in the natural language of the application and allowing the CAD system to distribute the required algorithms across an integrated network of programmable system components.

The successful supplier of components will have to offer more than a catalog with specifications of his parts. He must provide the hardware and software systems required for the systems designer to effectively apply the PSCs to his end application, including simulators, emulators, software development support and debug packages, test generators, and a hardware description language with verification programs for defining and checking logic array programmations. Also, he must provide a quick-turnaround, low-volume manufacturing capability to allow the developer to prototype, debug, and qualify his system before ordering production volumes. Semiconductor manufacturers' current inability or unwillingness to supply this service is a prime factor in the current rush by systems developers to establish or acquire an in-house IC design and production facility.

Obviously, the major VLSI semiconductor suppliers will be unable to supply this service to every system builder due to resource limitations. However, the entrepreneurial spirit is still very much alive in America's semiconductor industry. New companies will appear or existing companies will change direction to provide integration services to those system developers without the resources or volumes to do it themselves. They will buy large volumes of the basic logic arrays or unprogrammed microprocessors of their own or the manufacturers' design and develop and apply the applications programmation or interconnection for their customers.

Some suppliers and users of processor ICs propose the standardization of microprocessor architectures to avoid the high cost of redevelopment of the software and support systems and associated training and logistics costs, restricting performance improvements to those available through higher density and performance VLSI technology. However, I believe VLSI will provide the opportunity for new architectural approaches to processors and memories with high payoff in military systems. We must learn not to constrain innovation, but to effectively manage the application of innovative advances in software technology, which will eventually provide the answer to the search for efficient programming of the PSCs in the system designer's natural language.

C. The Systems

The application of submicron VLSI will not automatically provide instant solutions to the problems of LSI in military systems. Government and industry management must strive to reduce the barriers to the rapid integration of this new capability into our defense arsenal. The issues of milestone zero decisions, as mandated by OMB directive A-109, and NATO RSI (rationalization, standardization, and interoperability) are making program starts increasingly difficult. The use of new technology, particularly that for which no second source as yet exists, requires many manmonths of effort to justify and validate; it is often easier and quicker to drop back to our outdated technology, suffering the loss of the cost savings and improved performances. Perhaps the greatest gains from VLSI will be through the low-cost computers and memories that run the management systems that reduce the time required to rationalize, justify, and validate the use of VLSI in systems. However, even this will require changes in our standard procedures under which we in the military business operate.

We must provide the structure that allows for this technological insertion into existing military equipment and provide adequate rewards for the companies that do so effectively.

We have already discussed the effective force multiplication and life cycle cost reductions achievable with VLSI. The systems suppliers will establish their alliances with the semiconductor suppliers and they will bring forth new levels of system capabilities, but the new sophisticated systems will not be much smaller or much cheaper and the DoD's budget will not be reduced. The threats to our national defense are real and expanding. The new levels of integration must be used to make our complex systems even more reliable and capable or create new, innovative, low-cost weapons and command, control, communication, and support systems that can be purchased in large volumes to increase our effective ness. The key will be in the application of data processing to collect, re-duce, sort, and automatically respond to the massive amounts of informa-tion available to our systems and soldiers.

As we look forward to the first decade of VLSI, we are exceedingly hopeful of its impact on military systems. As we live and work through the decade, we shall find that the problems in the creation of VLSI de-vices and support will be frustratingly difficult, but VLSI will happen. Our attempts to integrate VLSI into military systems will be agonizingly slow and the impact will be less than we envisioned, but as we look back on the decade, we shall be amazed at both the progress and impact of the ad-vanced electronic capabilities.

ACKNOWLEDGMENT

The Staff at Texas Instruments.

Chapter 9

The Impact of VLSI
on the Automobile of Tomorrow

ALFRED J. STEIN

Semiconductor Group, Motorola Incorporated
Phoenix, Arizona

JOHN MARLEY
ROBIN MALLON

Motorola Semiconductors Japan, Ltd.
Tokyo, Japan

I. INTRODUCTION

Not many years ago any reference to new integrated circuits for the consumer market conjured up visions of expanding chip architecture for TV and radio applications. Today, this picture has faded, giving way to a much more pervasive one which extends the influence of electronics into "consumerism" to a degree that has long been predicted but has been slow in reaching fruition.

One area of consumerism that aptly demonstrates this changing picture is the automotive segment. For more than a decade forecasters have been emphasizing the susceptibility of the automobile to electronic control. Only within the last couple of model years, however, have we begun to see the implementation of these predictions in other than random applications in optional equipment. The picture is changing rapidly, however.

The approaching maturity of this potentially huge segment of the consumer electronics market is basically the result of four factors:

1. antipollution and fuel economy legislation in the U.S. and Japan;
2. the automobile manufacturers' growing confidence in the reliability of semiconductor devices under the adverse environmental surroundings of the automobile;
3. the evolution of the microprocessor as a viable instrument for achieving real-time control; and
4. the development of LSI and VLSI techniques that permit the implementation of computer capabilities at the low cost required for mass consumer applications.

II. HISTORICAL REVIEW OF AUTOMOTIVE ELECTRONICS

The marriage of automotive and electronics techniques had its beginning in 1929 with Paul Galvin's development of the automobile radio; and for three decades (except for the radio) the marriage was barren. This is hardly surprising in view of the natural incompatibility between the vacuum-tube-related electronics technology of the times and the low-cost, high-reliability requirements of automotive manufacturing, with its limited source of electrical power and its inherent space problems. It was not until the invention of the transistor and the development of a practical semiconductor (solid-state) technology in the early 1950s that electronics was recognized as a potential partner to mechanics in mass automotive applications.

Quite naturally the first manifestation of semiconductor applications appeared in the mid-1950s as a major modification of the car radio, where the replacement of tubes by transistors immediately eliminated the troublesome vibrator power supply, reduced power consumption and under-dash space requirements, and dramatically boosted the service intervals. Then, in 1960, with the development of suitable solid-state (semiconductors) rectifiers, electronics infiltrated the engine compartment of the automobile, where it precipitated a significant change in automotive engine accessories—a change from the well-entrenched dc generator to the smaller, less expensive, and more reliable alternator for the generation of the automobile's electrical power requirements.

This initial underhood infiltration of automobiles by semiconductor devices immediately led to widespread speculation of an imminent revolution in automotive technology; but subsequent progress was frustratingly slow, particularly in terms of the electronics industry's phenomenal succession of breakthroughs in other fields. Yet it is not surprising in view of the divergent and often incompatible philosophies of the two industries. Automotive manufacturers had already evolved a product that was highly reliable, mass producible at relatively low cost, and functionally acceptable to the mass market. Any major innovative changes, in order to be adopted, would have to significantly increase the sales appeal of the automobile while at the same time improving reliability and reducing cost. On the other hand, electronics technology, being fostered principally by the military and by the emerging computer market, was strongly stimulated to provide functional improvements, with cost cutting being only a secondary consideration. Thus, while electronics advocates could easily show potential performance improvements through mass adaptation of electronics technology, they had difficulty in convincing the automotive manufacturers that these improvements could be gained without sacrificing reliability, particularly under the harsh underhood environment of the automobile. Moreover, the greater cost of the electronics required to replace comparable, time-tested mechanical functions made the anticipated performance improvements a risky and expensive gamble. Subsequent efforts, therefore, concentrated on the use of solid-state technology to assist in overcoming some of the inherent problems of the mechanical assemblies rather than replacing these assemblies entirely.

One particularly vexing problem with mechanics was the relatively short life of the breaker points in the ignition system. Here, electronics was called upon not to replace the breaker points, but to increase their lifetime by reducing the current through the points—the major cause of the points' failure. This was accomplished through a development program that produced a special low-cost, high-voltage transistor that could interrupt high currents while being triggered from a very small current from the breaker points.

A similar problem was developing as a result of the continuous demand for greater engine power. This led to engines with higher compression ratios being run at higher speed. A limiting factor was the ability of the ignition coil to store enough energy fast enough for reliable ignition. Again, semiconductors were employed to enhance the mechanical deficiency, this time in the form of SCR (silicon-controlled rectifiers) switching circuits. These circuits stored energy electronically and then quickly transferred this stored energy, through the SCR, to the ignition coil. The result was high-energy ignition.

However, as the electronics industry gained experience in developing technologies adaptable to the uniquely harsh requirements of automotive applications, the auto industry's increasing confidence in the reliability of electronics led to the inevitable move toward replacing the more trouble-prone mechanical assemblies with equivalent electronics. First to feel the impact of this new thinking was the mechanical voltage regulator, which like the breaker points, was subject to early failure due to constant switching of high currents through mechanical contacts. Replacing the mechanical switches and relays with equivalent electronic switching circuits, using solid-state transistors and rectifiers, greatly reduced the failure rate.

To make electronic switching even more desirable, the electronics industry employed an assembly technique associated principally with military products, which resulted in substantial size reduction as well as improved reliability. This so-called hybrid circuit technique involved the use of unpackaged semiconductor chips, deposited on an insulating substrate (ceramic), and interconnected by means of wire bonds inside the final encapsulated device. (This was the forerunner of today's monolithic integrated circuits.)

The troublesome mechanical clock also succumbed to the higher-reliability promise of electronics through adaptation of highly stable precision oscillators already in use by the military and radio industry.

Subsequently, in a major breakthrough, electronics infiltrated the engine control function in the form of cruise control. Cruise control initiated the automobile into the rapid decision-making capability of digital logic. Subsequently, it also marked the debut of monolithic integrated circuits (primarily logic circuits borrowed from the rapidly advancing computer IC technology) in the automobile.

Throughout the 1960s the automotive industry flirted with electronics (semiconductor) manufacturers, expanding its use of discrete components in those functions where some penetration had already been made and holding out the promise of significant additional applications that would result in a potential avalanche of component requirements. Practically, however, subsequent progress continued to be slow. Throughout most of the 1970s, electronic penetration of the car consisted of more isolated peripheral functions, based largely on the proliferating progress in IC designs. A brief encounter with the controversial seat-belt interlock and the development of antiskid and antispin circuits used principally in large trucks and buses, etc., proved the practical decision-making capabilities of electronics but failed to produce the long-heralded breakthroughs. For semiconductors manufacturers, the carrot proved to be as elusive as in earlier years.

A. The Path to LSI

Even as the two industries searched for converging paths toward common goals, the solutions to the problems were evolving through the galloping progress of semiconductor technology. One of these was the exponential growth of component density within a single monolithic integrated circuit.

It began with individual semiconductor components (transistors, diodes, etc.), separately packaged and externally interconnected to form functional electronic circuits. As a small step forward, the semiconductor chips were interconnected in chip form on a ceramic substrate to form hybrid circuits—a complete functional circuit in the space previously required for a single component. Then, in the chip space required for a single transistor, manufacturers developed methods for depositing and interconnecting multiple components by chemical means. With each increase in component density came a corresponding improvement in circuit reliability and, eventually, a corresponding reduction of circuit cost.

Progress came rapidly. At first, a few components were used to form logic gates. Then tens of components were used to form monolithic flip-flops and operational amplifiers. Next came a hundred components as counters and other more complex circuits and a new term was coined, "large-scale integration" (LSI).

Quickly LSI became a moving target: a hundred gates per chip at the start, then a thousand, then tens of thousands, until, finally, even the term LSI depicted too small a quantity and the term VLSI (very large scale integration) began to appear. Today, single monolithic chips containing in excess of a hundred thousand transistor equivalents (as in large monolithic memories) are becoming commonplace and technocrats are beginning to anticipate a million components in a single chip of silicon the size of a small fingernail.

As one facet of the semiconductor industry continues to search for solutions to problems for ever-increasing chip densities, other technologists are beginning to look at this expanding capability as a "solution looking for a problem." Where, in the commonly accepted automotive architecture, would a single circuit with the complexity of thousands of transistors be applicable? The answer comes from the computer industry.

1. Enter the Microcomputer

The invention of the computer has influenced our society with an impact that can readily be compared with that of the industrial revolution in the 1800s. In the ultimate analysis, the results are even more dramatic, for where the industrial revolution only extended man's physical capabilities, the computer revolution conserves and extends brainpower as well. Although computers have already infiltrated virtually every major business and scientific endeavor, their penetration of modern society is still relatively small compared with the potentials that are beginning to emerge.

The emerging applications are not the result of improving or broadening a computer's capabilities. Large computers already have the speed, the capacity, and the versatility to handle the never-ending variety of tasks that might be demanded of them. Ironically, then, the anticipated explosion of computer control and communications applications is predicated on splitting the tremendous power of the large computer into smaller and more specifically apportioned capabilities, to bring them within economic reach of the public at large. In a word—*microprocessors*.

Microprocessors had been used as subsections of large computers and computer terminals for many years before they became a household word in the marketplace. Economics, however, prevented them from becoming widely accepted. Initially, where a microprocessor was needed in only small quantities, it was built on individual printed-circuit cards, using standard SSI/MSI logic circuits, demanding two to three hundred individual packages. Even with the lowest-cost TTL families, such cards would be priced far beyond the market for mass applications. On the other hand, where a large market was anticipated, as with electronic calculators, equipment manufacturers were joined by semiconductor manufacturers in the very expensive task of developing custom LSI chips. This reduced the assembly and hardware cost of the microcomputer but restricted its end use to specific, dedicated applications. Either way, the myriad of potential microprocessor applications had to await the development of a less expensive and more versatile line of applicable components.

The success of the electronic calculator did, however, pave the way for the new trend. It did so for two reasons. First, it showed that a circuit with the complexity of a microprocessor could be manufactured on a single chip of silicon. Second, it proved conclusively that such a single-chip LSI circuit, if manufactured for large-volume applications, could absorb the high development costs and still result in a very low cost end product. With these stimuli, the semiconductor industry embarked on a development program that yielded "universal" microprocessors with greatly improved capability and attractive pricing.

While the electronics industry, as a result of military and computer requirements, was developing techniques toward highly reliable LSI circuits and was, at the same time, struggling to find a truly pervasive penetration of the high-volume automobile market, the automotive manufacturers were wrestling with some problems of their own—primarily as a result of government involvement. This came about for two reasons: an increasing concern with ecology and the worldwide oil shortage. The result was legislation requiring the drastic reduction of engine exhaust emissions while demonstrating a progressive increase in fuel efficiency. The auto industry thus had some new words to add to its glossary of terms—"mandatory development" and "legislated invention."

The conventional automotive engine had not changed its basic form since before the turn of the century, and with nearly a century of progressive refinement, there were not many possibilities for quick improvements on such fundamental and continuing objectives. The result was a series of devices, "bolted" to the chassis, that treated emissions after they were generated and that improved fuel economy by basically decreasing performance—a not altogether satisfactory solution.

It quickly became apparent that truly effective solutions required complex engine management before the generation of emissions and the mechanical modifications alone were not the preferred solutions. Indeed, all proffered solutions defined the need for a computer of relatively high sophistication. At last, a match between an urgent need on the one hand and a viable potential solution on the other had been achieved.

a. Technical Developments. Early microprocessors were primarily designed for mathematical manipulation at speeds that were considered fast by human standards. When applied to an automotive engine, however, the speed was not fast enough and the ability to physically control real-time events became far more important than mathematical calculations.

Clearly, improvements had to be made in the microprocessors to enable them to operate at higher speeds and to give them the ability to perform control functions on mechanical assemblies. This caused the electronics industry to investigate more thoroughly the acquired automotive experience and to convert this into more complex devices appropriate for automotive use.

At the same time, armed with the capability of the microprocessor, the automotive industry began exploring other areas of vehicle control, safety, and creature comforts. Each area contained independent design goals, which quickly revealed the limitations of any individual type of microprocessor as a solution for all. The electronics industry was now hard-pressed to develop more powerful devices to keep up with the demand for more diverse flexibility.

It must be recognized that the automobile is basically a mechanical system, obeying mechanical laws. The microprocessor, an electronic system, performs according to a completely different set of rules. The physical gap between the two systems is filled by a group of components called sensors and actuators. These are electromechanical devices that convert mechanical motion into electricity and electricity back into mechanical motion.

Communication between the microprocessor and the sensors or actuators is handled through a series of often complex circuits, called interface circuits. Since automotive sensing assemblies are growing in sophistication, it follows that interface circuits must do likewise. The growing number and complexity of interface requirements complicate the development of a single, complete microprocessor that can economically handle all the variations and combinations of functions it might be called on to perform.

To service these various requirements adequately, a technique adapted from the computer industry was employed: that is, by developing a single, very powerful microprocessor and combining it with separate, independently designed interface circuits, a building-block approach was adopted that could handle the technically sophisticated demands of various automotive manufacturers. The price paid for this flexibility was an increased component count for an overall system, with a corresponding increase in assembly cost but a far smaller cost penalty than would be incurred by the development of a number of different microprocessors each dedicated to the performance of a different set of functions. Also, since automotive sensing assemblies are continuing to increase in scope and complexity, electronic interface technology must follow suit. Refinement and eventual standardization of these applications continue on a long chain of technological advances and new inventions.

III. PROBLEMS FACING AUTOMOTIVE ELECTRONICS

The auto industry has not been the cause of the invention of new technologies as much as it has been a catalyst for bringing new technologies together and expediting the development. It does so by representing a unique market; one that offers extremely high volume usage in return for products that solve the natural conflict among high technology, high reliability, and low cost.

For semiconductors, this is a particularly inviting carrot since there are few other industries whose product costs are as strongly dependent on production volume. Thus "automotive" not only stimulates the advancement of the state of the art but provides a volume subsidy to other semiconductor users who benefit from the cost reductions within these technologies that are mandatory prerequisites for automobile penetration. The quantum leaps made in the last decade in the field of integrated electronics are a case in point.

The tremendous strides that have already been made belie the complexity of the problems associated with a continuation of the progress in very large scale integration, particularly for the automotive industry.

A. High Technology and Costs

As an example of the problems involved in correlating high technology and cost, consider the case of the silicon wafers from which VLSI chips are produced. These wafers are (at a given time) of a certain size and must be sold at a certain price to maintain a desired profit level. For a practical component density that is achievable at a particular point in time, increasing the number of components on a chip increases the chip size and reduces the number of available chips per wafer. Hence, the price of each chip must be increased to maintain profitability.

Applying this situation to the computer industry, for example, creates no serious problem since the increased chip cost is more than justified by the overall reduction in the total number of chips required by this heavily electronic-oriented equipment. The automotive industry, however, is in the business of selling cars, and the electronics' cost is essentially an overhead expense. In the volume involved, an added penny of original cost could influence the acceptability of a component, and since far fewer electronic components are used in a car than in a computer, the cost increase may not be recoverable by an overall reduction in the number of components used. Therefore, progress involving cost increases may be unacceptable despite any functional improvement associated with a more complex chip.

Increasing component density by improving process technology (a continuing process) periodically permits an increase in component count per chip without increasing chip size. Yet, because of some inevitable flaws within the basic silicon wafer, increasing component density also reduces the yield per wafer because of the greater probability of such flaws appearing at sensitive portions of a number of chips. Until "experience" catches up with "capability," this could also cause an unacceptable price increase for the automotive industry.

B. A Problem with Heat

Increased component density means increased heat. Since a silicon chip can tolerate only so much heat, the maximum ambient temperature around the chip must be reduced to maintain reliable operation. This is in direct conflict with the continuing temperature rises in the engine compartment of a car, where increased mechanical congestion and higher engine temperatures are a continuous way of life. Again, new technological breakthroughs are required if progress in VLSI capability is to be translated into practical automotive applications.

C. The Interference Problems

In the limiting sense, an increase in on-chip component density is associated with a reduction in intercomponent spacing. This increases the chip's vulnerability to two types of disruptive interferences: rf injection and voltage transients.

Radio-frequency injection stems from signals from general broadcast transmitters and two-way communications units, with the vehicle's electrical wiring system serving as an antenna. This poses a threat of disrupting VLSI operation.

Protection against rf injection is primarily a shielding problem whose solution probably exists within the realm of experience already gained by the radio and communications industries. Voltage transients, however, pose a much greater threat.

The number of electromechanical assemblies in the automobile is on the increase. All these assemblies—air-conditioning clutches, electric motors, high-voltage ignition coils—have one thing in common: they all produce potentially destructive voltage transients on the vehicle's electrical wiring system—a threat that is increased as the density is increased. Today, VLSI devices are emerging with up to 64 pins on the interconnecting package. Each one of these serves as a potential entry point for the transient. Considering the fact that today's VLSI circuits may contain upward of 70,000 active components on a chip, any of which, if destroyed, can render the entire chip inoperative, the voltage transient problem must be seriously addressed if VLSI is to make serious inroads into the automobile.

D. The Digital/Analog Problem

VLSI is primarily a digital technology, whereas nearly all the stimuli in an automobile are analog in nature. For each of these inputs, an analog-to-digital and/or digital-to-analog converter is required. Obviously, it would be desirable to combine analog and digital circuitry on a single VLSI chip, but the two technologies have significantly different performance parameters. While some progress has been made toward marrying the two technologies in a single chip and while the outlook for combining the two disciplines looks very promising, present limitations still impose the building-block approach where separate A/D or D/A converters are employed as input–output interfaces between the sensors and the computer—a requirement that leads to increased cost and a reduction of overall reliability.

E. A Problem of Systems Definition

To date nearly all electronic systems developed for the automobile perform isolated and independent functions. Many of these independent functions, however, are stimulated by a common input (e.g., engine rpm) or activate a common output (e.g., warning light). This leads to redundant operations and, consequently, an unnecessary expense. One method for easing this situation is the design of an electrical system that permits time-sharing of the control functions.

F. High-Speed Low/Speed

In terms of computer speed, the automobile appears to be a slow-moving object indeed. Even the fastest engine functions, for example, are measured in terms of milliseconds, whereas computer operations in the submicrosecond range are easily achievable. A compatibility problem arises, however, when considering the tremendous number of functions that must be performed sequentially for the total monitor and control of the automobile with VLSI circuitry. This problem, in fact, makes an effective argument against VLSI and in favor of distributed (or effectively parallel) processing in automotive applications.

There are, however, a number of possible techniques that permit implementation of the potential cost advantage of VLSI while solving the timing problems. One of these, involving circuit design, increases throughput by adding internal functions such as clever timers, arithmetic multiplication and division, and direct memory operations to the VLSI chip. This would allow certain operations to be performed more rapidly than others or to run in parallel with other operations.

Another method is the increase of device cell speed and the subsequent increase in clock frequency through a reduction of cell geometry. The trade-off is the limit of how small the geometries can be made while maintaining acceptable yield and reliability.

Increased speed can also be obtained by driving the cells with more current and by reducing their resistance. However, greater current means greater power dissipation and a resulting undesirable increase in chip temperature.

Still another possibility lies in the realm of continuing developments in technology. Automotive VLSI proposals, to date, have involved N-channel MOS technology (NMOS) or the more recent HMOS process, because of their relatively high speed. The complementary MOS (CMOS) technology, despite its lower power dissipation and greater immunity to electrical noise, has been largely ignored because of its relatively slow speed and lower density. Currently, CMOS circuits using silicon-on-sapphire (SOS) technology have yielded speeds suitable for automotive applications but at an extremely high cost.

Obviously, the last word has not yet been uttered and the automotive industry still finds itself in a position to foster innovative developments that will swing the pendulum in favor of the most suitable techniques.

IV. The Potential Impact of VLSI on the Automobile

As world automotive manufacturers vie for increased market shares in the face of cost increases and energy shortages, automotive electronics offers an important new instrument for competitive, value-added performance features. Despite the differences in philosophies and problems in adaptation described earlier, implementation of electronics on a much greater scale than ever before is in the offing. A partial list of imminent features includes:

(1) human interface improvements,
(2) greater operating efficiency,

(3) remote control of power switches,
(4) time-sharing of information on fewer wires,
(5) internal and external diagnostics,
(6) dynamic ride control, and
(7) performance adaptability over operating life.

A. The Human Interface

New concepts in automotive signaling that provide both visual and audible alerts of existing or impending problems are ripe for early implementation. Warning displays that flash and buzz, that adapt their output levels automatically to changing ambient levels of light and sound, that will even provide physical prodding of the driver through seat mechanisms and talk to him directly through large storage of prerecorded speech have already been initiated. Like body styling, the imaginative use of audiovisual displays can be used to provide differentiation in the competition for the sale of cars.

The reaction of a vehicle to a driver's verbal command through speech interpretation via electronics has been discussed but is far more difficult to implement due to everpresent background noises and the legal implications involved in the event of misinterpretation of a command by the recognition system. Probably little can be done in the area from the standpoint of controlling a moving vehicle, although speech recognition could be used for entering simple monitor tasks while the vehicle is at rest.

B. Improving Automotive Efficiency

The continuous rise in the cost of energy, contrasted with the inexorable reduction of the cost for fixed-level electronic control, thanks to advances in and toward VLSI, makes electronic engine control one of the most promising and beneficial potentials in future automobiles. Although electronics has been considered an extra overhead expense in cars to date, the time is not too distant when electronic control will more than pay for itself in fuel savings over some predictable span of use. With the car's mechanical power train designed for optimum efficiency at a fixed set of conditions, appropriate sensors and actuators can be made to compensate for any variations in these conditions through the medium of the electronic (VLSI) brain, thus keeping the engine running at maximum efficiency.

C. Electronic Control of Power Switching

The concept of a "power bus" that locally supplies power to a clustered load in response to a low-power remote signal is not new. It is used in the aircraft industry to save weight and in industrial control to reduce power loss. When applied to the automobile, for weight and energy saving, the low-voltage (12 V) battery standard introduces the need for switching heavy currents and demands a very low switch impedance. Since on–off switching of automobile loads occurs at a relatively slow rate, a customized serial code can easily be generated within a microprocessor for the control of switches at a variety of remote and scattered locations within the auomobile. The successful application of remote load switching for such accessories as headlights, starter relays, horns, air-conditioning clutches, and the like must, however, awaits the successful development of a very low cost "intelligent power switch" circuit. Prices on the order of 10 to 20 cents per ampere have been postulated.

The power bus concept and an appropriate power switch diagram are schematically illustrated in Fig. 1. The switch diagram, however, is deceptively simple. In order to qualify, the switch must meet a difficult combination of specifications, including very high current gain, 1-mA turn-on sensitivity, very fast thermal overload shutdown, very slow (2 pps max) switching rates, and an off-leakage current of under 100 μA. Preventing oscillations in small-geometry, high-gain amplifiers, devising a very rapid thermal-rise detector, and integrating these functions in a very efficient, single-chip power switching circuit require a formidable combination of technologies. The successful implementation of this concept, however, must await the solutions to these problems.

D. Simplifying the Wiring Harness

A modern automobile may have upward of fifty actuators, relays, motors, and audiovisual devices scattered throughout the vehicle. A revolutionary simplification of the associated wiring harness could be achieved by imparting built-in intelligence and localized memory to these terminal assemblies. In this manner, the assemblies can be selectively energized by brief electronic messages sent serially along a single wire. Each "intelligent assembly" would respond only to the coded message destined for its designated location.

An example of this technique is illustrated in Fig. 2. Each "active" display is assumed to have built-in electronics permitting it to perform serial

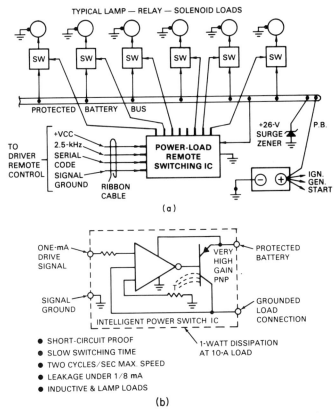

TYPICAL LAMP — RELAY — SOLENOID LOADS

(a)

- SHORT-CIRCUIT PROOF
- SLOW SWITCHING TIME
- TWO CYCLES/SEC MAX. SPEED
- LEAKAGE UNDER 1/8 mA
- INDUCTIVE & LAMP LOADS

(b)

Fig. 1. (a) The power bus concept and (b) a proposed switch appropriate for the application.

decoding, data retention, and display refresh functions. Therefore, a remote controller module can direct a constant stream of coded data along a single wire, with each segment of the data stream activating only the display for which it is specifically intended. The technique permits considerable space and weight savings in the instrument panel area and allows the designer to add or delete displays without significant redesign effort.

The flow rate of information from a remote display controller to such intelligent assemblies is very modest. An update rate of 10 to 12 substantiations/sec for each display signal should be adequate to keep the driver informed of even fast-changing data. The slow serial pulse rate that can be employed is easily filtered to screen out interference from electrical power transients.

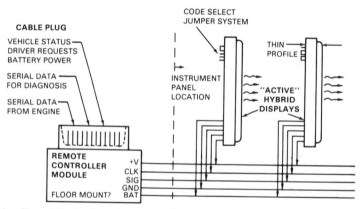

Fig. 2. Shared display bus. A common, five-wire ribbon supplies many separate display locations.

The diagram of a typical serial message structure that can service eight display panels is shown in Fig. 3. A single VLSI chip can organize information, choose from among several preplanned courses of action, and select from appropriate audiovisual outputs or machine control signals.

E. Automotive Diagnostics

Within the automobile, VLSI microcomputers can greatly enhance man–machine interface and provide operational control through rapid and precise manipulation of actuators and power devices. In addition,

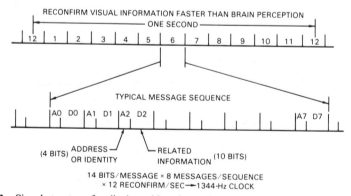

Fig. 3. Signal structure for display addressing. "Transmission," repeated 12 times/sec, can contain eight 14-bit messages, each destined for a different display.

they are also expected to play an important part in automotive performance analysis within the automobile itself and in garage diagnostics.

An envisioned automotive electronic control and display system is illustrated in Fig. 4. The system consists of three microcomputers, each assigned control responsibility for a major segment of the automobile's operation. One of these microcomputers (in this case, the entertainment and display center) is assigned the additional function of a "clearing house" for communication with all other internal microcomputers and with an external (garage) computer. Communications with the latter can be achieved with a simple five-wire (or fewer) ribbon cable. In this manner, the system can be put into a self-checking mode for pin-pointing problems via the built-in display and can interface with the external computer for more exhaustive troubleshooting. The latter, through appropriate serial command or data messages, can activate the various internal subsystems in a logical sequence for analyzing engine parameters, the ride control system, the entertainment system, and various other load terminations.

The concept of satellite control assemblies, each with a major assigned task but loosely tied together for information-sharing, should prove to be attractive to the auto industry. Since, traditionally, a vehicle is assembled in major subsections, each subsection could be independently analyzed by its dedicated microcomputer, with a final, overall test of these already qualified subsections occurring at the final assembly plant.

F. Dynamic Ride Control

A preferred method for reducing energy consumption of the automobile is to reduce its weight. This remedy, however, often leads to problems in cornering and braking and could adversely affect riding comfort over

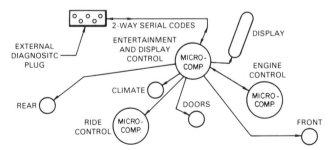

Fig. 4. A distributed microcomputer system that permits single-plug diagnostics from an external analyzer.

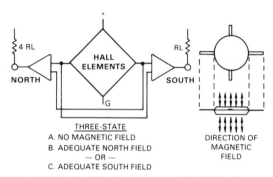

Fig. 5. Pellet Hall device detects presence and polarity of a magnetic field.

rough roads. These problems can be alleviated to some extent by dynamically changing the car's suspension characteristics to provide ride control.

Ride control requires a simple means for quickly determining changes in position or attitude of the vehicle with respect to a horizontal plane and quick compensation for any changes through strategically placed actuators. Rapid sensing, evaluation, and reaction, through computer control, are vital factors for success.

One method for determining movements and the rate of change of these movements is through strategically place sensors, perhaps related to the wheel shock absorbers. A possible sending device for this type of motion detection is a pellet-shaped, dual-polarity, magnetic field detector diagrammed in Fig. 5. Using a ground return and a single wire for supplying power and signals in combined form, this device can report the relative movement of coded magnetic patterns past one of its faces. The use of three-state loading, such as idle current for a neutral field, one unit of current for a north magnetic field, and four units of current for a south magnetic field, permits a single wire to carry all the information needed by a microcomputer to interpret magnetic field patterns.

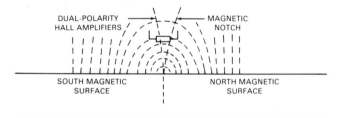

Fig. 6. Horizontal motion of detecting device with respect to a fixed magnetic pattern or movement of the magnetic pattern with respect to the detector results in a three-state output signal that defines their relative position and movement.

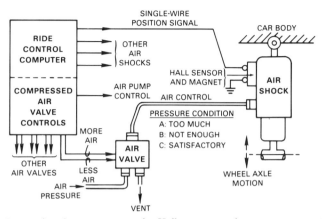

Fig. 7. Interaction between magnetic Hall sensor and computer-controlled, compressed-air shock absorber system offers rapid compensation for an otherwise rough ride.

The sensing principle is illustrated in Fig. 6. For any pattern of north and south poles on the surface of a magnetic material, there will always be a "neutral magnetic notch"—a region that has no magnetic field normal to the surface. This principle can be used to detect a wide variety of linear or rotational motions. The output signal, however, is not sensitive to the actual motion speed. In ride control applications, therefore, the VLSI microprocessor has the task of measuring the time patterns in this motion signal from one wheel and comparing the result with similar events from other wheels in order to determine how the automobile suspension should be adjusted for corrective action. Figure 7 depicts one possible method of ride control using compressed air and fast-acting air valves. Of course, these techniques of ride control also consume energy, so care is needed to assure that these methods produce a practical net savings of energy compared to a heavier automobile using totally mechanical ride control.

G. Adaptability of Performance

An interesting assignment to automotive electronics is the reporting of automotive component deterioration and routine maintenance requirements during the operating life of the vehicle. The owner of the automobile should be informed about the condition of tires, oil, battery, fan belts, and other finite-lifetime items. The lifetime is generally a mixture of operating time and elapsed time.

Other parts of the automobile are also subject to wear, although at a slower rate. In many such situations, changes in operating procedures can compensate for such wear conditions. In either case, however, some permanent record must be kept of elapsed time and operating time as well as of distance traveled. At certain predetermined intervals of time and travel distance, permanent records must be accumulated which cannot be altered by loss of power or human tampering. This historical "tally" can provide the following services:

(1) *Odometer.* Traveled miles in multiples of 25- to 50-mile units are reported. Finer details of distance do not need a permanent record.

(2) *Maintenance Warnings.* Periodic notices for oil changes, tire care, garage checkups, etc., are given.

(3) *Preplanned Behavior Changes.* The capability of VLSI microprocessor components to store alternative constants, subroutines, or entire substitute algorithms provides the automotive engineer with exciting opportunities to modify or adapt to age, mileage, operating hours, or elapsed time at some stressful condition of high rpm, overheated engine, malfunctioning pollution control, etc.

A microprocessor organization that can record historical events is shown in Fig. 8. Using VLSI techniques and an electrically "write-once" memory structure such as fuse-link programmable memory, much of this diagram could be integrated into a single-chip component for economic and security reasons. Like the flight log recorder in an aircraft, it should be possible to protect this component in the event of an accident and to read our various historical fact from the device.

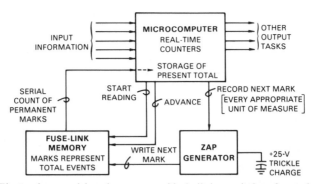

Fig. 8. Electronic record keeping can provide built-in reminders for periodic maintenance.

V. RUTS IN THE ROAD TO VLSI†

Penetration of the automobile by electronics technology in really significant proportions is critically dependent on two factors:

(1) the development of a high degree of cooperation and collaboration between the automotive and semiconductor industries, and

(2) the success achieved in the development of VLSI technology to meet the special requirements of the auto industry at the very low unit cost associated with automotive components.

Obviously, without the development of innovative automotive sensors and actuators, compatible with electronic practices, many of the examples described earlier will never reach fruition. Likewise, without ingenuity and creative flair in the definition and configuration of automotive VLSI circuits that can achieve high-volume, very low cost, semiconductor production, large-scale use of electronics will remain a long-term process, dominated by the one-at-a-time replacement philosophy of the past. Clearly, then, expansion of VLSI capabilities is the key technical criterion in the marriage of automobiles and electronics.

Many discussions of the future of integrated circuit technology—and hence of VLSI—begin with the chart in Fig. 9 showing the progress made by our industry in cramming increasing numbers of transistors onto a

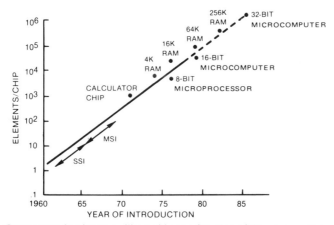

Fig. 9. Component density on a silicon chip as a function of time can predict progress in single-chip circuit complexity.

† See Stein [1].

single chip of silicon. This well-known curve agrees very well with the historical trends of integrated circuit technology—so well that the extrapolation of the curve is frequently taken as gospel by persons planning future technical developments. Extrapolation is an easy planning tool—all that is required is a ruler.

Extrapolation predicts that the most advanced integrated circuits in the 1985 time frame will have over one million transistors on a single chip, which is roughly equivalent to the number of logical gates comprising today's medium-to-large computer CPUs. Furthermore, this chip will require hundreds of interconnections and will operate at much higher speeds than today's LSI products.

This prediction, however, is subject to risks hidden in three specific areas: (1) market demand, (2) resources, and (3) technology. Each of these holds the potential of becoming a deep rut in the path to VLSI and may well lead to the conclusion that "extrapolation is not destiny."

A. The Market

All indications point to the continued expansion of the market for VLSI electronics. One explanation for this is an analogy describing the development of another key electrical technology product, the electric motor. In the late 1800s, industrial concerns typically had one large electric motor driving a series of belts and pulleys. To start his machine, an operator merely connected his belt to the nearest shaft.

In contrast, a typical modern home may have upward of fifty electric motors ranging from the air-conditioning and major appliance motors to those used in electric toothbrushes, can openers, cameras, calculators, etc. Developers of a wide range of products have discovered that the electric motor provides eagerly sought after product features.

In the same context, during the 1960s, industrial concerns typically had one large central computer. To use this machine, an operator merely connected his computer terminal, his card reader, or his printer to this single, central system.

Just as with the electric motor, product ingenuity will be the key factor in expanding the market for VLSI, as it too finds its way into a very wide range of products. The VLSI microprocessor (and other related integrated circuits) will be the "prime mover" for widely used electronic products of the 1980s, for which the automotive market is a prime candidate. The marketplace, therefore, does not appear to present a significant risk.

B. The Resources

The impact of advances in technology will be heaviest in the areas of capital and human resources. As technology advances, the cost of the more sophisticated equipment required is escalating. For example, in lithography, the capital cost per 1000 wafers/week of capacity is rapidly climbing (Fig. 10). A projected increase of two orders of magnitude in the cost of the equipment coupled with an order of magnitude decrease in the throughput results in a projected three orders of magnitude increase in the capital cost of lithography capacity.

An estimate of the capital cost of producing bits of memory for the last four generations is shown in Fig. 11. In spite of increased total capital requirements, the lithographic capital cost, per bit, has steadily dropped. This has happened because yields have improved, automation has taken hold, and densities have climbed. Thus there is the problem of increasingly expensive equipment used in the manufacture of steadily lower-cost products. The semiconductor industry, already a capital-intensive one, will become even more so in the future.

Human resources are also a major problem. A quick estimate, based on one company's sales per employee, is that the $2.8 billion U.S. merchant market for integrated circuits is based on the output of roughly 1500 circuit designers and is built on a process developed and maintained by 750 process engineers. By contrast, in 1978, the four major professional sports (football, baseball, basketball, and hockey) took in approximately $450 million, based on the efforts of a roughly equal number of major-league players. Yet our colleges and universities are producing candidates for professional sports careers at a far greater rate than they are candidates for integrated circuit design and process engineering careers.

The ability to attract and hold competent technical experts in the semiconductor industry will be a key to its future expansion. The productivity of engineers now in the industry must be greatly increased by design automation and standardization if it is to make the most efficient use of this most limited source.

SYSTEM	HOURLY THROUGHPUT	APPROX COST	CAPITAL PER 1000 WAF/WEEK
CONTACT	60	$ 15K	$ .03M
PROJECTION	60	$ 240K	$ .41M
DIRECT STEP	30	$ 480K	$ 1.63M
e BEAM	6	$1500K	$25.50M

Fig. 10.   Cost trends in lithography technology.

YEAR	MEMORY SIZE	DENSITY K BITS/in.^{2}	WAFER SIZE	RELATIVE CAPITAL COST
1973	1K	60	2 in.	1.0
1975	4K	195	3 in.	.21
1977	16K	585	3 in.	.09
1979	64K	1771	4 in.	.06

Fig. 11. Capital cost per bit as a function of time and component density.

Yet another limitation of resources is in the area of raw materials, particularly the availability of silicon. The chart in Fig. 12 illustrates the currently available world supply and the projected demand for polysilicon, measured in metric tons between 1973 and 1984. One easily sees part of a potential dilemma in that the apparent demand will cross over the projected supply in the 1982–1983 timeframe. This problem is due to the fact that today's semiconductor devices and solar energy products will both make increasing silicon demands throughout the 1980s. During this timeframe, unless new polysilicon capacity is added, this resource can become a serious stumbling block in the path of continuing expansion.

C. The Technology

The characteristics of the 1985 integrated circuit chips are shown in Fig. 13. By extrapolation, we expect to see 1 million transistors on a silicon chip 0.4 in. on a side. The most widely discussed technical problem is the lithographic challenge. Technical publications today are full of stories of exotic x-ray machines and electron-beam equipment that writes directly on a wafer. The minimum linewidths called for in 1985 range between 1 and 1.5 μm, which is within the optical capability of present technology.

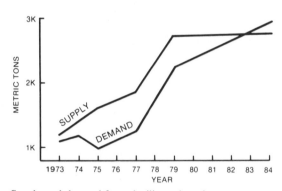

Fig. 12. Supply and demand for polysilicon, based on current capacity data.

YEAR	DEVICES PER CHIP	DIE SIZE (mils)	AREA (mils²) PER DEVICE	LINE WIDTH
1978	35,000	200 × 200	1.26	4 μm
1982	250,000	300 × 300	.36	2.1 μm
1985	1,000,000	400 × 400	.16	1.4 μm

Fig. 13. Forecast of component density.

A number of technical problems must be overcome in achieving the projected 1985 chip. For each of these problems, solutions are now available somewhere in the current integrated circuit technology. These solutions, however, still have to be adapted to VLSI. Take, for example, interconnections. Most current VLSI chips have a single layer of metal connecting circuits on the surface of the chip itself. This single layer occupies 30–60% of the area of a die and, clearly, is a major contributor to the size, hence the cost, of VLSI circuits. One solution, multilayer metallization, has been practiced for years in making bipolar digital TTL and ECL products. The technique is there; we need only to enhance it and adapt it to newer devices.

Other technical problems include the ability to etch small geometries and packaging.

The ability to produce the smaller geometries required for VLSI is influenced by the purity of the chemicals used in the etching process. When linewidths and spacing requirements reach the 2-μm level or less, as demanded by the envisioned density of 1 million elements/chip, impurity particulates in the etchants must be reduced in size to less than 1 μm. This could present a potential limitation in the quest for ever-increasing component densities.

In the development of new packaging for VLSI components, a number of assumptions will have to be made. These include such things as the fact that the number of leads may eventually be 250 or more, that random bonding pads or bumps will be located throughout the chip, that there will be 1-nsec speeds, that chip size will grow to 400 mils on a side or more, and that power dissipation requirements may exceed 10 W.

Based on those assumptions, a variety of new packaging techniques to resolve the problems has already been anticipated, including new die bonding processes and entirely new generations of equipment to assemble and to test these complicated new packages. These areas are being subjected to keen focus, and rapid progress is anticipated.

Then, too, there is the issue of inspection. Fully automated inspection techniques in the areas of design layout and drafting, mask making, and wafer processing will become critical in their ability to enhance and control yields during the manufacturing process.

Testing today's main-frame computer might require several days, but practical VLSI components will not have that luxury. To reach the required low cost level, the MPUs of the future will have to be tested in near real time and that will require whole new testing technologies and the development of chips that feature a self-testing capability. While some progress is being made, the state-of-the-art testing equipment is today too expensive and its throughput too low to satisfy future requirements.

All of the challenges just outlined in the areas of raw materials, equipment, capital, and technical resources really are not new to the semiconductor industry. All of them have been faced before and have been successfully overcome. In each case, the solutions to the problems of the 1985 integrated circuit have been demonstrated in the lab or are already in the process of implementation. We must only overcome the economic problems to adapt them successfully to the future requirements.

REFERENCE

1. A. J. Stein, *in* Morgan Stanley Electronics Letter, November 30, 1979. Morgan Stanley, New York.

Chapter 10

Very Large Scale Protection for VLSI?

HAROLD LEVINE

Sigalos & Levine
Dallas, Texas

Abraham Lincoln once said, "The Patent System added the fuel of interest to the fire of genius." This was earlier recognized by the founding fathers who, in the Constitution, gave Congress the power to protect an inventor and keep others from stealing his property and ideas, as an incentive to encourage invention and fulfill the constitutional mandate "to promote the progress of the useful arts."

I. A LOOK AT PATENTS—PAST AND PRESENT

A. Patents—Past

To gain a modern business perspective on patents, it is worthwhile to look at the historical development of the patent system. The early concepts of private property can be traced back to ancient Greek and Roman times and even earlier. In fact, the first patent in recorded history was granted in the city of Sybaris (an ancient Greek colony) in the year 500 B.C. for novel gourmet dishes, with a one-year exclusionary protection period. The avowed purpose of this monopoly, as recorded in the history books, was to induce others "to labour at excelling in such pursuits"; and, as is well known, the tradition of the rights of the individual, among them being the right to hold and defend private property, has been (and it is hoped, will continue to be) among the most important elements of the modern free society that we have inherited from those early times.

During the dark ages there seems to have been very little, if any, innovation; and, in fact, there appears even to have been a regression in technical progress. However, during the latter part of middle ages there began to be some recognition of the desirability of improving one's earthly lot. The customs of those times are only vaguely known, but there began to develop some organized procedures, though primitive, for handling technology and innovations and for according protection to inventors. For instance, in the Republic of Venice, beginning about A.D. 1300 officers were appointed to inspect canals, bridges, mills, and other engineering works. The kingdom of France had a royal council as early as 1330; and it was in this timeframe that the governments granted exclusive rights (usually in the form of royal decrees) to artisans and inventors for the practice of their innovations.

By the year 1500 inventions and patents flourished in the Italian states and the German Empire. Not long after, the custom of granting patents was adopted in the Netherlands and, shortly afterward, in England and France. Its introduction into England was accelerated by the flow into England of craftsmanship that was actively sought by the British Crown as a means to strengthen the Crown economically and militarily. By the year 1600, patents were firmly established in England and, subsequently, were extended into the American colonies. During the American Colonial period patents were frequently granted by a particular colony to its citizens.

It was not surprising, therefore, that the U.S. Constitution should contain a specific provision (Article I, Section 8) empowering Congress to grant protection for limited times to authors and inventors. This provision was implemented some 13 months after the 1789 adoption of the Constitution by "An Act To Promote The Useful Arts," which was signed by President George Washington on April 10, 1790. From these constitutional and legislative beginnings evolved the protection of the invention property rights in the United States as they are known today.

Since those early days of 1790, there have been a number of changes in the U.S. patent laws. However, the basic concept of providing an exclusive right to an inventor for a limited period of time has been steadfastly retained.

B. Patents—Present

1. What Is a Patent?

Consider, now, some fundamentals about what a modern patent is, what a patent does, and how a patent is obtained.

In the United States a patent is a grant that results from a contract by the inventor with the government. The inventor agrees to disclose his invention to the public and, in return, the government grants the inventor, for a limited period of time (17 years), the right to exclude others from making, using, or selling the claimed and protected invention. Thus what a patent does is to encourage public dissemination of new technology while protecting the exclusive rights of the inventor for a limited time. Note that the right is an exclusionary one; that is, the right to prevent others, not the right to the inventor himself to use his own invention. This distinction becomes important in situations where there may be preexisting generic patents owned by others that also cover the new invention.

2. How Are Patents Obtained?

Consider the business realities of the patent system as it now exists and as it relates to obtaining patents. First, it is appropriate to review the process by which patents are obtained. To protect an invention and obtain a patent, a patent application must be prepared. This is a complex set of legal documents usually prepared by a patent lawyer who, after approval by the inventor and client, forwards the application for patent to the Patent Office of the country involved; in the United States, it is the U.S. Patent and Trademark Office, where the application is examined. During the examination process there is usually an initial rejection of the claims for the invention; and the applicant, through his attorney, then files amendments and arguments for patentability. This may take several iterations and, ultimately, if the attorney is successful, a patent is granted with one or more claims defining and protecting the invention.

Patent systems in other countries may generally be considered to fall within either the pattern established in the so-called socialist countries, such as the USSR and its satellites, the Peoples Republic of China, and Albania, on the one hand, and the remainder of the world on the other. Under the socialist systems there are two principal avenues of rewarding or recognizing an inventor for his innovation. The first of these is called an *author's certificate*. Here, a state committee evaluates the worth of an invention and, in return for a complete assignment to the state of all rights, the state awards the inventor an amount of money based on the state's determination of the worth of the invention. The second course available to an inventor involves obtaining a patent. Such a patent theoretically carries with it the right to exclude others within that country from making, using, or selling the invention. However, since the means of production are all vested in the state and the inventor really does not have the option to prevent production by the state, he must search, through government bureaucracies, to find infringers and request from them appropriate compensation. Because this process is almost impossible to implement practically, statistics show that virtually no patents are taken out by citizens of the socialist countries; almost all follow the route of the author's certificate.

Turning now to the Free World systems, the patterns that exist there basically parallel those of the United States in granting for limited periods of time (usually 15–20 years), the right to the inventor to exclude others from making, using, or selling his invention. However, in direct contrast to the United States patent system, the laws of essentially all countries outside the United States contain compulsory licensing and working provisions that, simply stated, require that unless the invention is put into sufficient commercial use within a brief period of time (usually two or three years), so as to satisfy the reasonable commercial needs within the country, the government may grant to any interested person a compulsory license which permits the third party to practice the invention upon the payment of some reasonable compensation to the invention owner.

There is one additional major difference between patents in most countries of the world and those in the United States. Whereas once a patent is granted in the United States, under the present law there is no need to pay any annual taxes for maintenance of the patent, in essentially all other countries (except Canada), in order to maintain the patent in force, an annual payment must be made, which increases with the life of the patent and may become very substantial in the latter years of the life of the patent. Germany and Brazil, for example, have this system of increasing annual payments.

II. PATENTS—LICENSING AND LITIGATION

A patent is property and, as such, rights can be granted to third parties under the patent in the form of licenses to make, use, or sell the claimed and protected invention. Licenses can vary from nonexclusive to exclusive and each has its own implications. One important implication is the tax consequences. In the United States, for example, the patent owner in some cases, in a properly structured arrangement, may be able to obtain capital gains tax treatment for royalties or monies or other considerations he collects from granting an exclusive license or from selling the patent.

Very few patents in the U.S. are litigated. Many more are licensed and many are not used. For the patents that are litigated in this country, there is a very high mortality rate. The majority of patents are declared invalid in fully litigated U.S. proceedings. The patent mortality rate varies in the federal judiciary from circuit to circuit. The circuit most hospitable to patents in the United States is the Fifth Circuit, which includes Texas. According to a study of adjudicated patents on a circuit by circuit basis published by the Bureau of National Affairs, in the Fifth Circuit about 50% of the patents litigated were held valid while the other 50% were declared invalid. This figure for holding patents invalid ranges to as high as 87% in circuits such as the Ninth.

The cost of litigation to enforce patents in the U.S. is exceedingly high. A patent litigation can range anywhere from $100,000 to an incredible cost in excess of $1 million (in 1980 dollars), depending on such factors as the length, intensity, and complexity of discovery; the subject matter involved; the duration of the litigation; and appeals. Many times a litigant cannot control his litigation expenses. He has, in effect, an expense tiger by the tail. These litigation expenses are primarily for legal fees and disbursements and do not include the cost of the patent owner's time and involvement. For example, if the patent owner is a company, you would need to add to these estimates the cost of the company executives' time and the use of internal resources. One may readily and justifiably conclude the litigation is really a sorry way to settle patent disputes and should be used only when there is no other viable alternative and clearly only when the economics of the situation warrant and justify this level of expense.

The procedures for patent litigation abroad are significantly different and so are the costs—usually about one-third to one-tenth the cost in the U.S., depending on the country involved.

III. COSTS OF OBTAINING PATENTS AND PATENT ACQUISITION STRATEGIES

The cost of obtaining a patent in the United States from filing to issuance of the patent, including the government filing and issue fees and legal fees involved, can range anywhere from $2000 to $10,000, depending on the complexity and nature of the invention being protected. If the applicant wishes to file corresponding patents in foreign countries, the per country cost of filing presently ranges from $700 to $2000 and the total cost for protecting the patent on a multinational basis (including the U.S.) may run between $7000 and $20,000 on a typical six-country cumulative basis. This, of course, is a fairly sizable investment and it should be considered just that—an investment. This investment should be subjected to the same business tests as would any other investment.

Why should such an investment be made? How should such an investment be justified? It is important to examine the potential benefits to be obtained versus the cost of obtaining them and to examine critically the trade-offs. If the cost is not justified, clearly business judgment dictates that the investment should not be made.

There are many so-called experts today, primarily among the academic community and in government, who believe patents do not play a vital role in our economy. They believe, as Assistant Secretary Baruch has been quoted as saying, that "Patents are less and less an object to rely on." At a recent panel discussion at the government's Patent Lawyers' Symposium in Washington on "Federal Patent Policy and the Innovation Process," in which the author participated, one of the panelists was a member of the Office of Technology Assessment for the Congress of the United States. He felt that the trend in American business was away from patents and toward trade secrets and other forms of protection. It is this view that forecasts gloom and doom for the patent system. The opposite concept should be pursued.

The patent laws should be changed, certainly, but to strengthen the patent system so that the patent owner, the business community, and industry can rely on the validity of patents. The mortality rate of patents must go down, with a goal that 87% of the patents litigated will be held valid instead of invalid. This, of course, requires some sweeping changes to be made in the law in the U.S. Patent and Trademark Office.

IV. PATENTS AS A STIMULUS TO RESEARCH AND INVENTION

The business facts of life are that the patent system *does* stimulate research and invention. Very importantly, the system encourages investors to provide risk capital and resources to carry out extensive research and development projects.

For example, it took Merck and Co., 14 years to develop a drug for hypertension and congestive hear failure. DuPont spent ten years and $27,000,000 (in pre-World War II dollars) before nylon was produced. They also spent eight years and $25,000,000 for research and development before Orlon went into production. In the pharmaceutical field, of 5000 antibiotics discovered through research, only about 20 generally prove commercially useful. It is obvious then that research is both expensive and risky.

Alexander Graham Bell exhibited his telephone in 1876 at the Centennial Exhibition at Philadelphia, where it was considered a curiosity. Today the industry employs over 1,000,000 people. Hundreds of billions of dollars are invested in plants and equipment.

Xerox is a world leader in graphic communications. They employ over 100,000 people, with sales in excess of $5 billion/yr.

Why did these giants develop? Why was DuPont willing to invest so much time and money to develop nylon and Orlon? Why did Edison try over 1,000 filaments before he found the one that would work in a light bulb? Why did Chester Carlson (a former patent attorney) patiently go from manufacturer to manufacturer for over five years to get the Xerox process launched? There must have been some incentives, and there were. One of these incentives was the patent system. Obviously, hard-nosed, profit-seeking businessmen will hesitate to invest money in research and development facilities if they cannot expect that valuable inventive results will be legally protected by patents. On one occasion, W. G. Malcolm, chairman of the Cyanamid Company said, "The patent system is the life blood of Cyanamid's business. The same is true of almost all other major corporations in practically every industry."

Why is it the life blood? Simply because in the absence of a patent incentive, an attempt would be made to keep the inventions secret. It was this intense secrecy that characterized the medieval guilds and that undoubtedly retarded scientific progress and economic growth during the Dark Ages. It is interesting to note that for centuries the Chinese led the world in civilization and technological development. For instance, gunpowder, papermaking, and the compass were Chinese inventions. Yet the

Chinese have been technologically passed over by the world in the last two to three hundred years. One Chinese official said one reason for this was that China had no patent system, and when a man made a discovery or produced an invention, he tried to keep it a secret and passed it on from father to son in order to protect the idea and keep it a family matter. Compare that with the Forester patent on magnetic cores for computer memories which was a magnificent patent for producing income. It was reported by an electronics magazine that one computer manufacturer alone paid $12 million for the nonexclusive license rights to make, use, and sell the magnetic cores.

This, then, is one of the key thrusts of the patent system—incentive versus secrecy. It is to encourage time, effort, and expense for a reward, or as the early Greeks put it to "induce others to labour at excelling in (their) pursuits."

V. PATENTS AND THE SEMICONDUCTOR INDUSTRY—VLSI

The advent of the semiconductor industry with integrated circuits and, in particular, very large scale integrated circuits (VLSI) brings new problems and challenges to the area of protection of inventions.

The semiconductor industry in the United States is now a large industry involving a large number of corporations of varying sizes. From 1952 to 1968 alone, over 5,000 U.S. patents were issued on semiconductor devices [1]. It is safe to say that several thousand more patents on semiconductors were issued during the next decade. Further, the industry has proliferated new firms and, since 1971, the new firms have obtained about 50% of all semiconductor patents.

These new patents were obtained by massive expenditures and risk capital for research and development programs which brought about technological change at an extraordinary pace. As a result, it is not surprising to find that a firm with the latest technology may have production costs as much as 20 to 40% lower than comparable costs of those firms using technology obsolete for six months or more [1,p.64].

Technology is still growing at an explosive rate. As large-scale and very large scale integration appears in logic and memory circuitry, computer engineering is drastically changing. These large-scale integrated circuits are having a profound impact on the architecture of the computer by placing tens of thousands of devices on a single chip instead of multiple chips. The present aim is to produce chips containing over 100,000 devices each [2,3]. Thus the circuit designers are busy shrinking linewidth and length on current designs and increasing the complexity of logic circuits, layout topology, and circuit design. Such massive integration requires 1-μm lines and electron-beam lithography for implementation. This is required because the wavelength of light utilized in the state of the art processes of photolithography has become a significant percentage of the linewidth being printed. From the introduction of the 1K-bit dynamic RAM (random access memory using a three-transistor memory cell and P-channel metal–oxide–semiconductor (MOS) technology) to the recent marketing of 64K-bit dynamic RAMs using a single transistor cell and N-channel, double-level, polysilicon gate MOS, progress has been nothing short of dramatic [2]. There has been an unbroken downward trend in cell size and minimum chip gemoetries. Further, large chips containing from 100,000 to 500,000 gates make it possible to include processing power of a million instructions per second on a single chip [4].

Obviously such growth and innovation will require the investment of tremendous sums of risk capital. As an example, Honeywell will make VLSI its major internal circuit development thrust over the next three years and plans to spend several million dollars during that period for equipment and development costs [5]. Further, Gene Miles, director of memory components marketing for National Semiconductor Corporation, has been quoted, concerning the 1-megabit RAM expected to surface in 1985, as saying "that it will cost several million dollars to build that first chip" [3,p.115]. Thus it is apparent that such designs are not inexpensive and represent huge investments in their own right. Development costs of $100/gate are expected for a dense, complex state of the art design [4,p.47]. VLSI will place even heavier demands on the industry's computer-aided design capability. For the United States semiconductor industry, both government and commercial research and development funding may well be needed before VLSI becomes a reality. Large government-subsidized programs to reach this goal are already well underway in Europe and Japan [6–8]. If x-ray lithography is required to meet the density requirements of VLSI, it will require costly masks made by electron-beam patterning.

The costs of becoming involved in VLSI may be so great that only companies of substantial size and capability will be able to compete. Pierre Lamond, vice president and technical director at National Semiconductor Corporation, has been quoted as saying,

It's hard enough to get the investment capital just to stay involved, much less to start up. The industry will change considerably in the next few years, and there's one thing you can be sure of: there will be no start-ups in VLSI [3,p.122].

Further, such costs may require the VLSI technology to be shared within the industry. The importance of such sharing can be illustrated by the development of the semiconductor industry in the United States. The success of new firms in the industry in the early 1950s through the late 1960s was due, in large measure, to liberal licensing arrangements and the mobility of scientists and engineers [1,pp.77–78]. The liberal licensing standard was apparently set by AT&T and its subsidiaries, such as Bell Telephone Laboratories, which was the pioneer in semiconductors with the transistor, and the Western Electric Company. First, these companies realized that they could not keep such a revolutionary development secret. Second, they were concerned with an antitrust suit against them by the Justice Department in 1949. During the early 1950s, AT&T was the pioneer in, held the strategic patents for, and possessed know-how for the semiconductor industry. An apparent attempt to monpolize or even dominate the industry could easily have jeopardized their cause. Licenses were liberally granted on reasonable terms and conditions to all applicants. In this manner, the liberal licensing standard prevalent in the industry today was established.

Present patent licensing policies of the semiconductor industry seem to consist of either not licensing at all or nonexclusive licensing of patents in one of three ways listed in descending order of frequency in use:

(1) cross-licensing,
(2) lump-sum, paid-up licenses or annual lump-sum payments with a minimum payment required, and
(3) licenses on a running royalty basis.

Many of the companies in the semiconductor industry own patents or inventions that are widely used in the semiconductor art. Most of the large firms prefer licensing strategies that permit them to market their semiconductors free from patent interference.

The sharing of VLSI technology presents a different problem and challenge than the sharing of general semiconductor technology as we know it today. The semiconductor technology was basically patented, and thus licenses could be granted to competitors or other companies that desired them. The Japanese government-subsidized program was begun approximately four years ago and reports indicate that about 1,000 patents on VLSI technology have resulted [7,9]. It is presumed that these patents include those issued not only in Japan but also in the U.S. and other foreign countries. Such patented technology can obviously be shared through patent licensing programs.

The April 21, 1980, issue of the *Electronic News* [7,9] points out that $306 million was spent in the Japanese four-year government-subsidized

VLSI program. Of the 1,000 patents that have been applied for in this program, 59% are held by individual researchers, 16% by a number of participating companies, and the remaining 25% belong to research teams from the individual companies. A new $256 million five-year follow-on program in Japan to develop advanced computer operating systems, peripherals, and terminals will replace the VLSI project. While serving as the general patent counsel of Texas Instruments, in late 1974 and early 1975, I was forecasting what then appeared to be a concerning trend, namely, a shifting of the balance of patent power from the United States to abroad. Subsequent reports of Patent Office statistics for the years 1976, 1977, and 1978 seem to support the conclusion that U.S. industrial leaders in the semiconductor fields seem to be filing fewer patents while our competitors from overseas—Germany, Japan, France, England—seem to be significantly increasing their filings of patents in the semiconductor area.

Could it be that America is becoming less innovative as measured by the single criteria of patents? This is one plausible explanation, but I think not. I believe America is just as innovative today as it has been in the past and perhaps even more so. What we are seeing, I believe, is inadequate emphasis and recognition of the values of the patent system and the reluctance on the part of U.S. industry to make the investment necessary to protect the fruits of American industrial innovation. It is clear to me that such inadequate emphasis on patents on the part of American industry will be expensive in terms of sharply reduced future royalties from abroad. This tide must be turned around. A very vivid and current illustration of this concern is the thousand patents that have been filed as a result of the Japanese VLSI programs. American industry will one day have to reckon with these patents. Without a strong patent arsenal of its own, the only way American industry will be able to respond will be through payment of royalties.

VI. PROTECTION FOR VLSI TECHNOLOGY

Producing integrated circuits that contain hundreds of thousands of transistors involves development efforts in at least two main areas. The *first* is circuit design, including the physical layout, and the *second* involves fabrication, including lithography, preparation of the wafers, packaging, and testing.

Novel and inventive circuit designs can, of course, be protected by patents. The novel physical layouts required for VLSI sharply increase lithographic development problems. The trend is toward double and even triple or high multiple layers of polysilicon for new memory cell design [3,p.117;10]. Such processes create a physical layout that is increasingly more difficult to fabricate. The challenge is now to protect not only the lithographic process, but also the physical layout. Today there are no known meaningful patents on the topology of integrated circuits. Will copyrights protect topology? Will trade secrets provide adequate protection? What about those whose business it is to copy designs of others by making masks of existing devices and selling these masks to device copiers for as little as $150/mask? Do we need a new form of protection for circuit layout (or topology) development?

There are, at best, four methods that can be used to protect the new inventions developed during the creation of very large scale integrated circuits. These methods are: (1) patents, (2) copyrights, (3) trade secrets, and (4) unfair competition laws.

Patents, of course, can be used to protect compositions, articles, machines, and processes.

VII. VLSI PATENTS

Patents, for example, can be used to protect logic circuits and lithographic processes. However, the new VLSI circuits require complex topological layouts, including the configuration of registers, control circuitry, arithmetic and logic units, memories, and their interconnecting busses. It is very difficult, if not virtually impossible, to completely or comprehensively protect such topological layouts by means of conventional patent claiming. To develop a meaningful patent arsenal to provide adequate protection for the topology of complex VLSI circuits—aside from the questions of being able to achieve this with conventional patent claiming techniques—also presents a serious question as to the economic viability of such a strategy, noting that the cost of obtaining such patents may be inordinately high.

What is needed is an innovative, nonconventional approach to patent protection for VLSI topology which will provide effective protection at an economically viable cost. It must be recognized that pursuing a nonconventional or innovative patent strategy may intially require various appeals through the U.S. Patent and Trademark Office as well as the courts to establish the new patent strategy as being permitted by existing patent laws.

One such innovative strategy involves pursuing in the U.S. Patent and Trademark Office a claim involving a semiconductor chip wherein the topology is as shown in a particular figure of the drawings. It is common for chemical patents to be issued with claims drawn to compounds formed under certain conditions illustrated in a drawing of the patent application, and the claim that protects and defines the invention makes reference to that particular drawing figure. Such patent claims have been used very infrequently in relation to patents that have issued for mechanical or electrical devices. In 1941, the U.S. Patent Office Board of Appeals issued an opinion in Ex parte *Beyer and Tarn* [11]. The board considered a claim for a device for holding a fluid conduit member in position wherein a part of the claim specified the slope of a certain part of the device and the diameter of another part of the device as "being substantially within the calculation determinations of formula (13) of the specification." The U.S. patent examiner rejected the claims that made reference to formula (13) on the ground that limitations of a claim should be set forth in the claim itself rather than by reference to a formula to be found in the specification or in a chart shown on a drawing. The U.S. Patent Office Board of Appeals sustained the examiner's position.

In 1943, in In re *Faust* [12], a U.S. Patent Office supervisory examiner reversed the adverse ruling of a U.S. Patent Office examiner as to the form of claims containing three ingredients of respective specified percentage ranges with the relative percentages lying within areas defined by designated lines found in the drawing. The examiner rejected the claims as improper in form, stating that they should be complete in themselves and not depend on the drawings. The examiner did not raise any question as to the intended scope or definiteness of the claims and thus the question was simply one of policy on the part of the U.S. Patent Office as to whether an applicant may have recourse to that form of claiming in defining his invention. The supervisory examiner reviewed the history of the U.S. Patent Office policy and stated that the long-established rule that a claim should be self-contained and should be verbally stated should not be departed from lightly. He pointed out that it was a well-accepted practice that the parts of a claimed structure must be defined in words rather than by referring to the drawing by the use of reference characters for the disclosure of such subject matter (*Ex parte* Osborne, 1900 OG 137). He pointed out that the first material departure from the rule that each claim must be complete within itself came with the decision (*Ex parte* Brown, 1917 CD 23) in which the dependent type of claim, that is, one claim referring for a portion of the subject matter to a preceding claim, was approved. The supervisory examiner, in reversing the U.S. Patent examiner and approving the applicant's claim format, stated that he saw no reason why a like holding should not apply where one portion of the subject matter appears in the printed part of the application and the other in the drawing. The supervisory examiner pointed out that long-established usage requires that in the ordinary case the claim be stated in words alone and, wherever practical, this usage should be observed. However, he also stated that in situations such as the present case, where, according to the petitioner, the verbiage necessary to define the areas now defined by reference to the diagram was so extensive and involved as to defeat the very purpose of a claim, an exception to the usual rule should be made.

Probably the most famous case in this area is Ex parte *Squires* [13]. Squires' claim 1 reads "A font of numerals as shown in FIG. 1." Claim 1 was rejected for failing to particularly point out and distinctly claim the subject matter. It was not certain whether the U.S. patent examiner rejected the claims because they were not stated in words or because of the reference to the drawing. The applicant for the patent offered the substitution of the language in claim 1 with FIG. 1 of the drawing presented physically within the claims. On appeal, the applicant (appellant) pointed out to the U.S. Patent Office Board of Appeals numerous instances where reference in the claim had been permitted for ranges and/or tabular material in either the specification or drawings in order to avoid awkward reproduction thereof as part of the claim language per se. The appellant pointed out there is no statutory requirement as to the precise form of claiming. The board, in holding for the appellant, stated that the applicant did not claim broadening phrases such as "substantially" or "substantially as described" and felt that it was impossible to suppose that anyone who wished to determine the scope of claims 1 and 2 would have any difficulty in identifying what the claims covered. Thus the board was of the opinion that under the circumstances of the particular case reference to a pictorial showing of the outline of the object sought to be covered by the patent is the best, most accurate manner of defining the invention. The board saw no reason why representation of FIG. 1 should be repeated in the body of the claim. The board pointed out that it contains no extraneous matter that would tend to detract from definiteness. Thus claims 1 and 2, which referred to the drawing, were allowed.

It will be noted from these cases that there is *no* statutory requirement setting forth a particular manner of claiming an invention except that the claim point out and distinctly claim the subject matter that applicant regards as his invention [14]. Further, it will be noted that reference to a pictorial showing of the outline of the object sought to be covered by the patent claim may be the best, most accurate way of defining the invention which depends for utility on the precise configuration of the objects and their interrelation in a set. It seems to me that a convincing argument could be made that such a description applies to the topology of integrated circuit chips and, in particular, should apply to the unique and complex topology of very large scale integrated circuit chips. As stated earlier, it would be extremely cumbersome, if not practically impossible, to claim the precise arrangement of the elements forming the topology of a VLSI chip, noting that the verbiage necessary to define the topology would be so extensive and involved as to defeat the very purpose of the claim. Thus I believe that (as set forth in *Faust,* supra) a convincing argument can be made for an exception to the usual rule to provide for protection in this very important and growing area of circuit topology.

VIII. VLSI COPYRIGHTS

Under present copyright law and present interpretations, copyrights do not appear to offer much encouragement as a mechanism for protecting VLSI circuit topology. When one considers the significant investment in time, personnel, and resources to develop complex layouts for circuits such as a 64 K RAM, it is almost incomprehensible that adequate protection does not exist today for these tremendous investments. Although I believe that the copyright law should protect circuit layouts against copying by others, the recent case law, as it is developing, seems to be pointing in a direction against such protection.

For example, in the recent case of *Data Cash Systems Inc.* v. *JS&A Group Inc. et al.,* 203 USPQ 735 (1979) [16], an action was brought for copyright infringement and unfair competition against the defendant because the defendant allegedly reproduced, imported, distributed, sold, marketed, and advertised copies of the plaintiff's computer program. The defendant produced a chess computer having a ROM that was identical to the ROM in the plaintiff's chess computer game. The defendant's ROM was actually the plaintiff's copyrighted computer object program embodied as a mechanical device. The court held that the defendant's mechanical device (i.e., the ROM) did not constitute a copy of the plaintiff's copyrighted program. The court further noted that plaintiff's ROM (i.e., the mechanical device) did not itself have copyright protection and that no copyright notice appeared on the plaintiff's ROM. Further, the court quoted Nimmer [16a, §2.18[F]] with approval to the effect that "Mechanical devices which cannot qualify as pictorial, graphic or sculptural works are not writings and therefore may not obtain copyright protection." Although this case was heard in the United States District Court for the Northern District of Illinois, it appears to illustrate the state of the law regarding the copyright protection of a mechanical device or an article that does not otherwise qualify as a pictorial, graphic, or sculptural work. Copyright law in its present state of development does not appear to be a viable alternative for the protection of VLSI topology.

IX. VLSI TRADE SECRET PROTECTION

Trade secret protection does not seem to be an attractive alternative for the protection of VLSI topology. Such units must be sold and when they are sold, the trade secret is lost and the concepts under trade secret law become part of the public domain. It would be well within the ability of those skilled in the art to perform reverse engineering to discover the manner and form in which the particular topology is achieved.

There are those who quite properly feel that trade secret laws form the best protection for their investments in software [15]. This is so because, under present law, software per se cannot be patented and can be protected by copyright only in its readable form, not in its mechanical form [16]. Thus software, at least in its readable form, could be protected by either copyright or trade secret. The owner can control knowledge or use through contractual agreements when trade secrets are used. However, when the mechanical embodiment of the software (e.g., the ROM chip) is sold, the trade secret protection is lost to those who can reverse engineer or duplicate the item. Therefore, a different and more effective method of protection must be sought.

X. THE LAW OF UNFAIR COMPETITION

Protection of novel or original work to create topology for VLSI circuits may be available through utilization of unfair competition law.

The landmark case in the field is the 1918 U.S. Supreme Court case of *International News Service (INS)* v. *Associated Press,* 248 U.S. 215, 63 L.ED. 211, 39 S.CT. 68 (1918). In that famous case, the plaintiff and defendant were both engaged in the business of gathering and distributing news. The defendant, by purchasing early editions of newspapers published by the plaintiff, was able to use the information in its own business. The defendant argued that all property rights in the uncopyrighted news were lost after publication and dedication to the public and could be used by anyone for any purpose. The Supreme Court decided the question solely on the theory of unfair competition. The court stated that simply because the defendant had the right as a purchaser of a newspaper to communicate the news set forth in that paper freely to anyone for any purpose he could not transmit the news for commercial use in competition with the plaintiff. The court said that the defendant admitted taking news articles that had been acquired by the plaintiff as the result of organization and expenditure of labor, skill, and money and that could be sold by the plaintiff and, thus, the defendant was endeavoring to reap where it had not sown by appropriating and selling the news as its own. The Court said,

> Stripped of all disguises, the process amounts to an unauthorized interference with the normal operation of complainant's legitimate business precisely at the point where the profit is to be reaped, in order to divert a material portion of the profit from those who have earned it to those who have not; with special advantage to defendant in the competition because of the fact that it is not burdened with any part of the expense of gathering the news. The transaction speaks for itself and a Court of equity ought not to hesitate long in characterizing it as unfair competition in business (p. 239).

The case is not without problems from the standpoint of precedent in its present-day application to VLSI protection. There was a strong dissent by Mr. Justice Brandeis who stated that, as a general rule of law,

> The noblest of human productions—become, after voluntary communication to others, free as the air to common use.

Probably because of the strong dissent, the courts, until recently, have not frequently followed this case or tried to encourage the growth of this doctrine. I believe, however, that the INS doctrine could be applied to protect chip topology. There is a direct and clear analogy between the facts in the INS case and the case where a competitor simply copies by making or buying a mask or set of masks for reproduction, a VLSI chip originated by his competitor.

In today's world of ever-increasing complexity of circuits with increasing demands for higher *densities,* today's designers, with the assistance of computer-aided design, will continue to innovate new circuit layouts with greater densities and efficiencies, all at tremendous investments of risk capital. There are companies in existence today whose primary activity is to copy and reverse engineer circuits by making and selling masks to competitors to produce designs that they did not originate.

Chief Justice Hughes once stated that the concept of unfair competition should apply to the misappropriation of what equitably belongs to a competitor [17]. Such misappropriation is a form of unjust enrichment and the INS case actually expresses the law's opposition to unjust enrichment. The courts, however, until recently, have tended to strictly construe and limit the INS case and doctrine to the facts in the INS case. In the case of *Raenore Novelties* v. *Superb Stitching Company,* 47 NYS 2d 831, 61

USPQ 59, (1944), the defendants copied a closure originated and designed by the Plaintiff's assignor. The plaintiff urged that the recent court decisions expanded the doctrine of unfair competition to include the act of copying an article or piracy within the area of forbidden trade practices. In support of its position, the plaintiff cited the INS case. The court stated that it had been referred to no pertinent authority that had held that, in the absence of patents or copyrights, a property right may be acquired in the functional features of an article or design so as to prevent its appropriation by others. The court said that, on the contrary, the publication of a device or a design unprotected by patents or copyrights projects its functional characteristics into the public domain and renders its public property.

However, this doctrine and attitude appear to be changing according to Callmann, a well-known and respected author, who states that the INS case postulated a new concept of unfair competition in focusing upon the competitive relationship and stressing the reciprocal rights and duties that are peculiar to and emanate from it. He states that it is clear that a competitor may be properly enjoined from doing what a noncompetitor may be permitted to do [18], Sect 60.3, p. 509. The law is basically opposed to unjust enrichment; i.e., "one who reaps where he has not sown." Thus Callmann states that the Supreme Court, in the INS case, was opposed not to the appropriation of the news as such, but to an appropriation of the plaintiff's business organization such that the defendant's conduct amounted to an unauthorized interference with the normal operation of complaintant's legitimate business [18, p. 510]. The Court, therefore, properly invoked the theory of unjust enrichment and disregarded the copyright concept with respect to the news. This rationale, of course, is directly applicable to VLSI topology.

It has been said that the well-known Sears and Compco U.S. Supreme Court cases have overruled the INS case. In these cases, the Supreme Court pronounced that when an article is unprotected by a patent or a copyright, state law may not forbid others from copying that article because forbidding copying would interfere with the federal policy of allowing free access to copy whatever the federal patent and copyright laws leave in the public domain. However, Callmann [18, Sect. 60.4(c)] responds to this pronouncement by quoting the opinion in *Schulenburg* v. *Signatrol, Inc.,* 50 Ill. App. 2d 402, 200 NE 2d 615 (1964):

> The shortcut used by defendants enabled them to compete with plaintiffs on equal terms by sparing them the necessity of spending time and money in acquiring the "know-how." Competition is a desideratum in our economic system but it ceases to serve an economic

good when it becomes unfair. The concept of fair play should not be shunted aside on the theory that competition in any form serves the general good. Only fair competition does that. Unfair competition is not competition at all in the truest sense of the word.

Callmann [18, Sect. 60.4(c)] is quoted in the recent case of *Data Cash Systems, Inc.* v. *JS&A Group, Inc.* [16]. This case, as discussed earlier, was an action for copyright infringement and unfair competition brought by the creator of a copyrighted computer program. As discussed, the court dismissed the copyright claim on the ground that the ROM is not a copy of plaintiff's computer program and therefore the copying is not actionable. Thus the court confirmed that copyright protection is available for a computer program in readable form but indicated that no such protection is available when the copyrighted program is implemented in a noncopyrighted mechanical device. However, the Court refused to dismiss the unfair competition claim stating,

> While the rule may once have been otherwise, unfair competition is not confined to the passing off of the goods of one for those of another and the "palming off theory" is no longer the sole and exclusive criterion in determining the right to relief against unfair competition. There may be unfair competition by misappropriation as well as by misrepresentation, that is, the doctrine of unfair competition has been extended to permit the granting of relief in cases where there was no fraud on the public but where one, for commercial advantage, has misappropriated the benefit or property right of another and has exploited a competitor's business values.

The Court further took pains to point out that Sears and Compco have *not* overruled the INS case.

Other recent cases have held similarly. Thus, in California, the Court of Appeals, Second Appellate Division [19], held that the duplication of recorded performances without authorization in order to resell them for a profit presents a classic example of the unfair business practice of misappropriation of the valuable efforts of another. The court further stated that such conduct unquestionably constitutes unfair competition even if there is no element of "passing off" (or "palming off").

In New York [20], the court held that the contemporary New York unfair competition law encompassed a broad range of unfair trade practices generally described as the misappropriation of the skill, expenditures, and labors of another.

The Sixth Circuit [21] has recently also upheld the common law right against unfair competition in a case involving "pirated" or "bootlegged" tapes that were reproductions of the original recordings marketed under a different label. The court cited the Sears, Compco and INS cases, supra, for support.

Thus we see the courts returning to the INS doctrine of misappropriation as constituting unfair competition. Inasmuch as protection of VLSI chip topology may be exceedingly difficult, if not impossible, as a practical matter under the present patent, copyright, or trade secret laws and inasmuch as that topology will undoubtedly be strikingly unique, complex in design, and highly expensive to develop, I believe that the misappropriation doctrine should apply if, after the circuits are sold, a competitor should copy them and then sell them for profit. No meaningful distinction can be seen to exist between this case and the record pirating cases.

XI. CONCLUSION, SUMMARY, AND GUIDELINES FOR THE FUTURE

In the state of development of the law today, there is no clear-cut, well-defined, reliable form of protection for topological developments in VLSI, and as such, we really do *not* have very large scale protection (VLSP) for VLSI.

Patents under the present laws and systems in the United States can provide very large scale protection for VLSI but only for selected aspects such as new circuit components, new circuits (including cell design) and, in some cases, the dimensional relationships between circuit elements in the very densely populated chips, fabrication and mask-making techniques (e.g., photolithographic, electron-beam, and laser techniques), chemical compounds, and chemical techniques. However, barring an innovative, bold, new thrust in presenting and claiming VLSI circuit topology, such as that represented by the "picture claiming" strategy previously discussed, there really is no effective or practical means of protecting VLSI circuit topology under the patent laws and system of the United States as they exist today.

Although some well-known companies are reported to be attempting to secure protection for VLSI and chip topology through copyrights by including a copyright notice on the masks that are reproduced on the chip itself, the present state of the case law seems to be pointing in a direction away from affording protection for topology under the copyright laws.

Trade secrets, as discussed in detail in this chapter, really do not offer any practically reliable avenue of protection.

The law of unfair competition presents a bright ray of encouragement for protection of VLSI circuit topology against unauthorized competitive copying. I believe that we will see much more litigation in this area as designers and contributors to the art attempt to prevent the unfair usurpation of the fruits of their investment by those competitors who merely copy and compete.

A. Guidelines for the Future

Until a new, special form of protection for the important area of VLSI topology is developed, I would propose the following strategy to be followed by those organizations that risk capital investment to develop new VLSI circuits and that wish to optimize their protection.

1. Patents

Aggressively pursue obtaining patents (on a selective but comprehensive basis) for all new developments and improvements in the areas of new circuit designs, cell designs, interconnections of circuits, maskmaking techniques, fabrication techniques, lithography, processing, testing, and design automation. For circuit topology, taking into account market expectations and the prospect of copying by competitors, patent protection should be vigorously pursued using the ''picture-claiming'' technique described in this chapter.

2. Copyrights

Although the present copyright law and its recent interpretations do not seem to encourage the prospect of copyright coverage for VLSI circuit topology, I nevertheless recommend the practice of applying a copyright notice to the masks of original circuit topology designs with the notice appearing on the resulting chips. This will preserve the option to enforce your copyright in the event the law becomes more favorable. Furthermore, having a copyright may help in enforcing rights under the law of unfair competition, noting that throughout the entire development of Anglo–Saxon jurisprudence, the courts, by and large, have found unjust enrichment to be morally repugnant to society and have gone to great lengths in many cases to prevent such injustice.

Harold Levine

REFERENCES

1. J. E. Tilton, "International Diffusion of Technology," p. 57. The Brookings Institution, 1971.
2. R. N. Gossen, Jr., The 64-Kbit RAM: A prelude to VLSI, *IEEE Spectrum, pp.* 42–45, (March, 1979).
3. R. P. Capece, Tackling the very large-scale problems of VLSI: A special report, *Electronics,* pp. 111, 114 (November 23, 1978).
4. G. H. Heilmeier, Needed: A "miracle slice" for VLSI fabrication, *IEEE Spectrum,* p. 45 (March, 1979).
5. S. Gross, Honeywell to spend $5M in 3-year VLSI project, *Electron. News,* (Nov. 5, 1979). p. 55
6. 100,000+ gates on a chip: Mastering the minutia, *IEEE Spectrum,* p. 42 (March, 1979).
7. J. Hataye, Japan to share VLSI Technology, *Electron News,* p. 78 (Jan. 14, 1980).
8. J. Smith, Common Market eyes VLSI effort, *Electronics,* pp. 92, 93. (Sept. 28, 1978).
9. *Electron. News* (April 21, 1980).
10. R. Capece *et al.,* Devices meeting focuses on VLSI, GaAs, and sensors, *Electronics,* pp. 129–133 (Dec. 7, 1978).
11. 51 USPQ 331, 332 (1941).
12. 86 USPQ 114 (1943).
13. *Ex parte* Squires, 133 USPQ 598, 599 (1961).
14. 35 United States Code (USC), Section 112.
15. Call trade secret laws best way to protect investments, *Electron. News,* p. 54 (Jan. 14, 1980).
16. Data Cash Systems, Inc. v. JS&A Group, Inc., 203 USPQ 735 (1979).
16a. M. B. Nimmer, "Cases and Materials on Copyright and Other Aspects of Law Pertaining to Literary, Musical and Artistic Works," 2nd ed. West Pub., Chicago, 1979.
17. A.L.A. Schecter Poultry Corp. v. United States 295 US 495, 522, 55 S.Ct. 837 (1935).
18. R. Callmann, "The Law of Unfair Competition, Trademarks and Monopolies," 3rd ed., (1968).
19. A & M Records, Inc. v. Heilman, 198 USPQ 425, California Court of Appeal, Second Appellate District, Div. 3 (1977).
20. Ideal Toy Corporation v. Kenner Products Division of General Mills Fun Group, Inc. *et al.,* 197 USPQ 738, (D.C. SD. N.Y.) (1977).
21. A & M Records, Inc. *et al.* v. M.M.C. Distributing Corp. *et al.,* 197 USPQ 598 (CA 6) (1978).

Index